“A splendid study of [illegible] ples of the world's greatest modern thinker . . . a strong p[illegible] the human nature and the great achievements of a sublime individual.”

—*Booklist*

“A solid work . . . explains how his life affected his science . . . topics like electromagnetism become graspable, and as our comprehension increases, so does our appreciation.” —*San Diego Union-Tribune*

“A fascinating tour of Einstein's great life.”

—*Minneapolis Star-Tribune*

“A lucid and fast-paced biography on an extraordinary life.”

—*Nashville Book Page*

“Colorful . . . new and interesting details . . . accurate, accessible explanations of theory.” —*Kirkus Reviews*

“A deftly crafted biography.” —*Publishers Weekly*

“An absorbing and lucid account of the public man, and a most comprehensible summary of theories . . . and much else.”

—*Sunday Express* [London]

“Mind-stretching but lucid, the three-dimensional models and analogies offer a rare adventure in enlightenment.”

—*The Sunday Times* [London]

MICHAEL WHITE is Director of Scientific Studies at d'Overbroeck's College, Oxford. He has contributed numerous articles to science journals, has scripted and hosted notable science programs on radio and TV, and written several science biographies for young people. He lives in Oxford, England. JOHN GRIBBIN received his Ph.D. in astrophysics from the University of Cambridge before becoming a full-time science writer. He has been a staff member on leading science journals, has written science articles for *The Times*, *The Guardian*, and *The Independent*, and has authored more than fifty books. He lives in East Sussex, England.

EINSTEIN
A Life in Science

Michael White
and
John Gribbin

A PLUME BOOK

PLUME
Published by the Penguin Group
Penguin Books USA Inc., 375 Hudson Street,
New York, New York 10014, U.S.A.
Penguin Books Ltd, 27 Wrights Lane, London W8 5TZ, England
Penguin Books Australia Ltd, Ringwood, Victoria, Australia
Penguin Books Canada Ltd, 10 Alcorn Avenue,
Toronto, Ontario, Canada M4V 3B2
Penguin Books (N.Z.) Ltd, 182–190 Wairau Road,
Auckland 10, New Zealand

Penguin Books Ltd, Registered Offices:
Harmondsworth, Middlesex, England

Published by Plume, an imprint of Dutton Signet,
a division of Penguin Books USA Inc.
Previously published in a Dutton edition.

First Plume Printing, March, 1995
10 9 8 7 6 5 4 3 2 1

The Library of Congress has catalogued the hardcover as follows:

White, Michael.
Einstein : a life in science / Michael White and John Gribbin.
p. cm.
ISBN 0-525-93750-1 (hc.)
ISBN 0-452-27146-0 (pbk.)
1. Einstein, Albert, 1879–1955. 2. Physicists—Biography.
I. Gribbin, John R. II. Title.
QC16.E5W47 1994
530'.092—dc20 93-42626
[B] CIP

Printed in the United States of America

Contents

Preface

Since this book first appeared, other writers have made allegations about Albert Einstein's life and character which do not so much suggest that the idol has feet of clay, but imply that it is entirely made of clay. This is surprising, not least because those allegations were made largely on the basis of the same documentary material available to us while we were writing our biography, and we found no evidence for such an extreme view. It has been claimed, for example, that Einstein was a womaniser who revelled in the attentions of intellectual groupies, that he treated his children badly, and that he was solely to blame for the divorce from his first wife, who ought to have been given credit for helping him to develop the special theory of relativity.

All of these stories contain an element of truth, and all are addressed in the present book. But the suggestion that these aspects of his character dominated Einstein's entire life, or are of more than passing interest, is absurd. Yes, Einstein did enjoy the attention of women, as most men do; yes, he was a distant father who did not have a close relationship with his children; yes, by putting his work first he did precipitate the divorce (but gave his ex-wife, Mileva, the financial proceeds of his Nobel Prize, on which she could live securely); yes, Mileva did help develop the special theory of relativity, but only by checking Albert's arithmetic.

Throughout the twentieth century, Einstein has been perceived as not only a great scientist but also a great human being. The most attractive aspect of the man is that he was not a cold, inhuman figure bent over a mass of scribbled calculations—there were many other facets to his personality. He was a man greatly preoccupied with politics, religion, philosophy, and the human condition. He was musical, widely read, and interested in all aspects of life.

Because of the esoteric nature of Einstein's theories, he quickly gained a reputation as a godly man and was increasingly identified with his work, almost as an actor becomes synonymous, in the eyes of the public, with their stage or screen role. Because of this, people began to see the great scientist as somehow other than human, a man of pure intellect, unable to feel or behave as others do.

As the examples of Elvis Presley and John Lennon have shown, it is the fate of twentieth-century icons to have the minutiae of their lives raked over in the search for sensation after they are dead. Einstein was certainly a twentieth-century icon, but his image has been protected from this attempted demythologising until now largely because his papers were in the care of an over-protective group of his literary heirs. Now that those papers are more freely available, both serious scholars and sensation seekers have access to those papers and are able to make their own interpretations of what they contain.

Our interpretation of the evidence is that Einstein was neither God nor Devil, but an ordinary man, with mortal weaknesses, and an outstanding talent for deciphering the mysteries of the Universe. The most important and interesting thing about Einstein remains his scientific work, which is the main focus of our book. The interest in his private life seems, to us, to concern his education and the political turmoils of the time, which saw him move around the world in search of the peace and tranquility he needed in order to carry out that work. If you want to know how often Einstein brushed his teeth, you will not find the answer in our book. But if you want to know how he came up with the theory of gravity, or proved why the sky is blue, or worked out the physics of a cup of tea, and if you want to know why a confirmed pacifist decided, after much soul-searching, to urge the development of the atomic bomb, then you will find the answers in these pages. We hope that with this account, the reader will see all the faces of Albert Einstein, a man as complex in personality as the theories he gave the world.

—Michael White and John Gribbin, September 1993

CHAPTER ONE

A Fascination with Science

When, in the early hours of 18 April 1955, Albert Einstein died at the Princeton Hospital, New Jersey, his brain was removed and kept for medical research. Although this dubious honour was not unique to Einstein, the mere fact that the great man's brain should be thought a subject of research at all shows the level to which his fame as a scientific genius had grown. Extensive study unearthed nothing special, but that is beside the point; though Einstein's brain was normal in every respect, the intelligence which had resided there for seventy-six years certainly was not.

His international celebrity ranged far beyond the scientific community. The French singer Edith Piaf always kept on her bedside table a picture of St Theresa, a Bible and a book about relativity. When she arrived in New York in 1947 and was asked whom she would like to meet first, her reply was 'Albert Einstein'.

To many, Albert Einstein was the archetypal absent-minded professor, a white-haired, eccentric scientist whom the uninitiated saw as the role model for mad professors everywhere. Although he was first exposed to public awareness at the age of forty, it was as the elder statesman of science, the Nobel laureate, peace campaigner and Zionist sympathiser that he was caricatured and metamorphosed into an icon. By public confusion, he was also somehow transformed into the father of the atomic bomb.

The reason why a little-known European physicist should

have, overnight, been propelled into the spotlight has puzzled commentators ever since. Somehow, Einstein became the twentieth-century personification of the wise man, the possessor of secret knowledge. It only enhanced Einstein's semi-mythical status that he happened to emerge on to the world stage in the winter of 1919, immediately after the worst blood-letting the modern world had known, with a conceptual framework which transcended the world and its problems.

Einstein was much more than a great physicist. To most people he was an undisputed genius. The reason for this was simple. Einstein's theory of relativity amounted to a completely new way of looking at the universe. The scientific community were seen to hold his theories in awe, and nobody else seemed to understand them; ergo, in a more innocent world, the man was a genius.

As a Jew living in an age of genocide, Einstein was actively Zionist; not simply anti-Nazi, but anti-German. He was a talented musician, deeply philosophical and polymathic. He was a man full of contradictions, whose beliefs and views on many subjects showed a surprising fluidity. An extremely likeable, although often distant, personality, he had many friends, but few close ones. Twice married, he was the father of two boys. Yet he needed solitude and mental isolation to work efficiently. He was definitely no team player.

The city of Ulm is situated on the left bank of the Danube at the foot of the Swabian Alps and lies almost equidistant from Stuttgart, some sixty miles to the northwest, and Munich, to the southeast. It was here, at Bahnhofstrasse 135, that Albert Einstein was born on 14 March 1879.

The German Empire, together with the rest of western Europe, was still recovering from the turmoil of the Franco-Prussian War of 1870–71 and the worldwide financial crash of 1873. Otto von Bismarck was German chancellor and the country had begun to enter a phase of great industrial expansion. The year of Einstein's birth also marked the coining by Wilhelm Marr of the term 'anti-Semitism' and the foundation of his League of Anti-Semites.

The second half of the nineteenth century was an era of increasing monopolisation of German industry by giant conglomerates and the start of a trend of depopulation in rural areas. The rural Jewish population of southern Germany fell by 70 per cent between 1870 and 1900.

Both Einstein's paternal and maternal grandparents were contributors to this sociological change. Albert's father Hermann Einstein had arrived in Ulm in 1866, aged nineteen. His family had made a move of some thirty miles north from the small town of Buchau, where the Einsteins had lived since the 1750s. Hermann worked for a cousin's company, becoming a partner in the firm some time around the year of his marriage, 1876. He had a natural bent for mathematics and had shown great promise at school.

However, Albert's grandfather Abraham Einstein did not have the financial means to pay for Hermann's further education, and his academic ability was left untapped. Abraham Einstein died in the prime of his life and Albert never knew him. Hermann Einstein met his future bride Pauline Koch in Cannstadt, a small town some miles away from Ulm.

Thanks to the success of the grain business run by Pauline's father and uncle in Cannstadt, the Kochs were wealthier than the Einsteins. They were also somewhat unusual in that both brothers and their families lived in the same house, the wives sharing domestic chores as the brothers shared business responsibilities.

Pauline and Hermann were married on 8 August 1876 in the synagogue in Cannstadt. The bride was seventeen. Despite an age difference of eleven years, by all accounts the relationship was a very close one and their marriage extremely happy.

In his parents can be seen the two early influences which produced the adult Albert Einstein. Hermann was a contemplative man with a clear mathematical mind. He was pedantic and always studied a situation from every angle before committing himself. He was also a very gentle man who found it difficult to say no to anyone. These personality traits were not only to appear strongly in the emotional make-up of his son, but were also to lead the Einstein family into severe business troubles a few years later.

Einstein's mother was a tall, slim woman who enjoyed robust health for most of her life. Although not considered beautiful, she had radiant grey eyes which shone with what has been described as a 'waggish twinkle' and expressed an earthy, native wit. She was brighter than her husband and more cultured. She was well read and was particularly interested in music. It was she who introduced Albert to the piano when he was a small boy and suggested that the children should receive musical instruction at an early age.

For the first year of the Einstein marriage the couple lived in Buchau where Hermann Einstein ran his own business. By 1877, however, they were back in Ulm, moving to premises on the Münsterplatz. Then, with Pauline six months pregnant, they moved again at the end of 1878 to number 135 City Division B, a flat in a drab, four-storey building which in 1880 was renamed Bahnhofstrasse 20 and later destroyed in an Allied bombing raid in 1944.

With financial assistance from the Kochs, Pauline Einstein's parents, Hermann was able to set up an electrical-engineering business in a building a few hundred yards away from their home on the south side of Cathedral Square. Pauline Einstein's side of the family were to lend their financial support to Hermann and his various business ventures on a number of occasions in future years and, as fortune would have it, they saw very little capital return on their investments.

It was in the flat on Bahnhofstrasse that Albert Einstein was born just before noon on a sunny Friday, 14 March 1879. The couple registered their son's birth the following day. His birth certificate, number 224, reads:

> Before the undersigned registrar appeared today, in person known, the merchant Hermann Einstein residing in Ulm Bahnhofstrasse B No. 135 of Israelite religion, and reported that to Pauline Einstein nee Koch his wedded wife of Israelite religion who lives with him in Ulm in his residence on the fourteenth of March of the year one thousand eight hundred seventy-nine in the morning at half past eleven o'clock a child of male sex has been born, which was given the first name Albert. Read, approved and signed, Hermann Einstein. Registrar Hartmann.

Family legend has it that when Einstein's mother first saw her son she thought that she had given birth to a deformed child because the back of the baby's head was extremely large and angular in shape. A doctor soon reassured her that there was little to fear. Nonetheless, even as an adult Einstein did have an unusually large head. Einstein's maternal grandmother was also disturbed by the sight of the infant because in her opinion he looked too heavy. Her first reaction upon seeing the baby was to throw up her hands in surprise and exclaim: 'Much too fat! Much too fat!'

According to family accounts, Albert was a very quiet and contented baby who needed little care. However, once fears for the shape of his head had passed, the family grew concerned that he was not developing at a normal pace. He was slow to learn to talk, and then developed the peculiar habit of constructing whole sentences before uttering a word. He could be seen to recite the sentence silently, his lips moving, before he came out with anything. He appears to have been a rather introverted and shy child. There is nothing in his childhood to signal that this slow-thinking and initially backward merchant's son would become one of the most distinguished thinkers in history.

During the first year of Einstein's life, his father's financial situation was going from bad to worse. Hermann Einstein was simply not cut out for business. He was too good-natured to be competitive and too indecisive to put the brakes on and make the correct moves when things started to go wrong.

At some point during the first half of 1880, Hermann's younger brother Jakob suggested that the two of them should set up in business together. Jakob had studied engineering and was keen to start an electrical-engineering and plumbing business. There was just one thing he needed to get started – money.

The deal was that Hermann would become a partner and secure funding from his in-laws, the wealthy Kochs. Jakob had all the ideas and inventiveness, Hermann the experience in business. It seemed like an ideal time to set up such an enterprise just as the world was converting to electric lighting, and the brothers started off with high hopes.

Munich, the capital of Bavaria, with a population of more than a quarter of a million, was a city easily large enough, the brothers had decided, to support a new business in the field of electrical engineering. The move was made in the early summer and the Einsteins registered as citizens on 21 June 1880. Munich would be Einstein's home throughout his childhood, the family remaining there for the next fourteen years.

At first the firm flourished and was even starting to show a profit, but then near-disaster struck. The ambitious and overconfident Jakob had invented a dynamo which he believed had a huge market. The problem was again financial. The manufacture of the dynamo on a large scale required a bigger plant, expensive machinery and labour. With Hermann's help, the funding for the project was secured from the extended family, the lion's share again coming from the ever-helpful Kochs.

Whether Jakob had grossly misjudged the market or the dynamo was not so great an invention as he had believed is not known, but the project did not succeed as they had hoped. The business remained on the edge of financial ruin for a number of years, propped up by constant injections of cash from the family. Small engineering works were finding it increasingly difficult to compete with giant companies such as Siemens and Halske, Union AEG and Kummer.

Thus, Albert grew up in an environment which, although happy and emotionally sound, had an underlying insecurity due to the precarious nature of the family finances.

In November 1881, Einstein's sister Maria, always to be known as Maja, was born. She was Albert's closest childhood friend, and brother and sister maintained a deep love for one another until the latter's death shortly before her seventieth birthday in 1951. Under her married name of Maja Winteler-Einstein, she wrote a short history of her brother's early life.

From this family account we learn of Albert's early ability to arbitrate in arguments and that he did not join in the games of the other children but instead 'occupied himself with quieter things'.[1] Much of this type of anecdote, found to litter biographical studies of Einstein, may be taken with a pinch of salt. It is very easy, even for a much-loved sister, to glamorise events in hindsight. An otherwise normal four-year-old child who was quiet and contemplative and kept himself to himself would probably be nearer the mark.

In Maja's account we also hear that Albert had inherited a bad temper from his maternal grandfather, Julius Koch, and that he would occasionally throw a temper tantrum. 'At such moments his face would turn completely yellow, the tip of his nose snow white, and he was no longer in control of himself,' his sister recounts.[2] This is perhaps significant only given the extreme gentleness for which Einstein was famous as an elderly man. Such behaviour is far from uncommon in young children.

We are fortunate to have Maja's reminiscences of her brother's early life because Einstein himself was particularly bad at recollecting his personal past and he rarely spoke about his youth.

Einstein's inability to remember details from his past was all the more surprising because his memory for scientific matters never failed him. He could easily recall details of things that greatly interested him many years after their first discovery, but could not recollect many events in his personal life. He once told a family doctor, Janos Plesch, that he found the phenomenon quite astounding and that it was perhaps a matter for psychoanalysts to look into.

Indeed, the eminent psychologist Anthony Storr has a great deal to say on this topic. He believes that this could perhaps be seen as supporting the notion that Einstein demonstrated schizophrenic tendencies and that Einstein's early determination that he would free himself from all personal ties is a common trait in schizophrenic characters. Einstein's extreme distaste for authority, demonstrated in his rebellious stance at school, his desire to remain stateless and his unequivocal hatred for the German nation as well as his detachment from the personal by his total lack of interest in clothes or creature comforts, may all be seen as supporting this mental state. The fact that Einstein had a very poor memory for his own childhood demonstrates a subconscious attempt to eradicate a personal history and further detach himself from the real world. In fact, Storr, in his fascinating study *The Dynamics of Creation*, goes as far as to suggest that if Einstein had not been schizophrenic he could not have developed his theory of relativity because its creation could only emerge from a personality with a strong sense of detachment, a mind that did not want to identify with a physical presence and could stand back and observe from outside.[3]

Einstein himself once suggested that he had been able to develop his theories because as a child he had been intellectually retarded and therefore only turned to such matters as space and time in adulthood. Most adults, he believed, would not consider such things important. He apparently believed that his intellectual achievements were based on a type of naivety or innocence which is lost at some point by most people as they make the transition from child to adult.

The Einsteins lived first in a small rented house in Munich but with the initial success of the electrical-engineering business, after five years in the city they were able to move to a much larger property in the suburb of Sendling. This was a large, rambling house set in its own grounds, surrounded by trees. It had a huge, unkempt garden in which the Einstein children played. Hermann and Jakob's tiny factory was within walking distance and for a time business was progressing well. However, Albert and Maja lost their garden when, in 1894, the family business again failed and the estate had to be sold to a property developer.

The years in Munich were rich in experiences for young Albert. Against the fashion of the time, his parents encouraged their children to gain early a sense of independence. At the age of three or four he was sent through the busy streets of Munich. The first time he was shown the way, the second he was allowed to make the journey alone but observed, the third without any supervision. According to the recollections of his family, Albert was a very obedient child who had an unswerving respect for his parents. He observed their instructions to the letter and stopped at every street, looked left and right and crossed streets with great care. Maja has suggested that their parent's insistence on giving the children self-reliance ingrained in Einstein's character the need for emotional independence in later life and enabled him to look after himself confidently when the need arose, though this did not extend to domestic responsibilities.

At the age of five, Albert Einstein was enrolled at a Catholic elementary school (*Volksschule*). The Einsteins were not practising Jews and the academic standards of the school were of far greater importance to them than its religion. The fact that the local Jewish school was a considerable distance from their home and that the fees were higher than at the local Catholic school convinced them that the Catholic *Volksschule* would be best.

According to family history, Einstein was not especially happy at the elementary school and was not noted for his brilliance by the teachers there. When Hermann Einstein asked the advice of the headmaster regarding the best future career for Albert, the headmaster said that it didn't matter because Albert would never make a success of anything.

Einstein's difficulty seems to have stemmed, even at the age of seven or eight, from his unorthodoxy. He was not the type of pupil who took the standard route through a problem and was often slow to formulate a solution. He was not willing unthinkingly to accept the teacher's every word. He showed no particular aptitude for mathematics, although, according to his sister, what he lacked in speed and accuracy he made good by being reliable and persevering in his work.

The school was strict and the teaching highly formal. Rote learning was the accepted method of the day and the use of *Tatzen*, or knuckle-raps with a cane, was considered a legitimate way of instilling multiplication tables into the minds of children. The narrow-minded approach of the teachers at the *Volksschule* is illustrated by a now famous incident which Einstein recalled many years later to a friend. During a religious-studies lesson, the teacher produced a large nail and declared: 'The nails with which Christ was nailed to the cross looked like this.'

Much has been made of the cruelty of this statement by Einstein's biographers, who assumed that, with Einstein being the only Jewish student in the class, the teacher went on to point out that the crucifixion was carried out by Jews. There is no evidence that this did in fact occur, or that the incident brought home to the schoolboy that he was different in any way from the rest of his classmates. Einstein appears to have been unaware of his Jewishness until at least his late teens.

At home, he continued to find enjoyment in his own company.

He was happy to play with Maja, but he was most content when absorbed in some puzzle or other. Like the boy Isaac Newton, he liked constructing models, his favourite toy being a building-block set. He also built houses of cards, sometimes as high as fourteen storeys. Einstein's skill with such things demonstrated an early strength of determination and perseverance.

At the same time in Einstein's life there was a counterpoint to his more analytical interests. This came from the influence of his mother and, through her, his growing interest in music. Albert began violin lessons when he was six years old. At first he was taught music by rote and, according to his sister, this almost spoiled things for him. As at school, any form of regimented learning aroused his resentment and hindered his progress. However, he soon realised the need for tireless practice and became very fond of music.

This interest stayed with him for the rest of his life. By the age of thirteen he had begun to realise the beauty of Mozart's music and took constant pleasure in performing it. He was never a very accomplished musician, more of an enthusiastic amateur. His violin playing was later seen by many as a safety valve for the mental strain of working on his scientific theories, and, after Einstein had achieved international fame, his violin became a kind of trademark.

According to legend, Einstein was set on the road of scientific enquiry at the age of five. Albert was ill in bed and to cheer him up his father showed him a pocket compass. The boy became very excited by the fact that whichever way the compass was turned, the needle remained pointing north. It was his first realisation that a totally enclosed object could be influenced from a distance by an invisible, untouchable force. Einstein later said:

> That this needle behaved in such a determined way did not at all fit into the kind of occurrences that could find a place in the unconscious world of concepts. I can still remember – or at least believe I can remember – that this experience made a deep and lasting impression upon me. Something deeply hidden had to be behind things. What man sees before him from infancy causes no reaction of this kind; he is not surprised by the falling of bodies, by

> wind and rain, nor by the moon, nor by the fact that the moon does not fall down, nor by the differences between living and non-living matter.[4]

At the age of ten, Einstein left the Catholic elementary school and was enrolled at the Luitpold Gymnasium. This was an even more formal establishment. Although Albert had always found regimented learning difficult to cope with, he had at least obeyed his teachers and gained good reports throughout his elementary-school days. He had even been at the top of the class, according to a letter from Pauline Einstein to Albert's grandmother Fanny, dated 1 August 1886: 'Yesterday Albert got his grades, once again he was ranked first, he got a splendid report card.'

However, before long, Albert began to clash with the system at the gymnasium. Many of the teachers were fixed in their ways and showed little imagination in the way they taught their subject. With personalities like Einstein's they appear to have been heavy-handed and impatient. Lacking any sympathy for the way his mind worked, they immediately decided that he was something of a wastrel. In addition to this was the fact that the school, like many others of the time, showed a heavy bias towards the arts and the humanities rather than the sciences. Both Greek and Latin were high on the agenda, and although Einstein enjoyed the logical rigour of the latter, he never did get to grips with Greek and often aroused the anger of the Greek master, who, echoing the words of Einstein's old headmaster, declared that nothing would ever become of him.

Ironically, when the school, situated at Müllerstrasse 33, was totally destroyed in a World War II bombing raid and rebuilt on a different site, it was renamed the Albert Einstein Gymnasium.

There was one teacher at the gymnasium, a Herr Reuss, whose approach differed from the rest in that he believed that it was essential to make the students think for themselves. In this way he managed to generate in both Einstein and his other students a deep fascination with ancient civilisations. When Einstein had achieved worldwide recognition and was himself a university professor, he decided, on the spur of the moment, to pay his old teacher a visit. Einstein arrived at the teacher's home dressed in his usual casual clothes and was astonished to find that the elderly Reuss, oblivious of his old student's international fame, refused to see him. Insult was then added to injury when, thinking that Albert Einstein was a beggar looking for a hand-out, Reuss had his maid send him packing.

Despite its educational failings, Einstein gained a great deal from his time at the gymnasium. He also realised that he would have to work on his own and develop a programme of self-education. A great influence in this matter was Uncle Jakob, his father's business partner. Jakob gave him a little book about algebra, which instantly caught the boy's imagination. 'Algebra is a merry science,' Jakob said. 'We go hunting for a little animal whose name we don't know, so we call it x. When we bag our game we pounce on it and give it its right name.'

Later, at about the age of twelve, Albert decided, quite independently, that he was going to prepare himself for future work at school by spending his summer getting to grips with more advanced algebra. Jakob would set him problems and was often amazed at the thoroughness of his nephew's work and his ability to find an answer to the most difficult calculations the engineer could devise. Albert even managed to find a totally original proof for the Pythagorean theorem during this summer vacation.[5]

There was obviously no shortage of adults ready to make their contributions to Einstein's early intellectual development. With his father and uncle on one side with their engineering and scientific influence, and on the other the more artistic nature of his mother, he had the good fortune of receiving a balanced education above and beyond his formal education at school.

The year 1889 was a significant one in Einstein's extracurricular education, thanks to his first encounter with a twenty-one-year-old medical student called Max Talmud (he later changed his name to Talmey after moving to the United States).

It was customary at the time for middle-class Jewish families to help poorer members of the community who were studying and had limited means of support. Talmud was studying medicine at Munich University and was introduced to the Einstein family through his brother, who was a practising doctor in the city. He visited the Einstein home every Thursday evening for supper and, despite the difference in their ages, he became a great friend to Albert.

In return for his weekly supper, Max Talmud would bring along popular science books to lend to Albert. In this way Einstein was introduced to such works as A. Bernstein's *Popular Books on Physical Science*, *Kosmos* by Alexander von Humboldt and L. Buchner's *Force and Matter*, which formed the basis of his early scientific education and took him far beyond the level of study he was receiving at the gymnasium. The most important book Talmud passed on to his young friend was a popular geometry textbook called *Lehrbuch der ebenen Geometrie* by Spieker. According to Einstein's own statements later in life, it was the single most important factor in turning him on to science and mathematics at an impressionable age; he had just turned twelve. He also later referred to this textbook as the 'holy geometry book'. From that point on, science became an integral part of his life.

Years later Talmud wrote a book about relativity and Einstein's early life called *The Relativity Theory Simplified and the Formative Years of its Inventor*, which was first published in 1932. In his account he states that Einstein's mathematical ability was so prodigious that before long he had outstripped Talmud's knowledge of the subject.

When Einstein was thirteen, Talmud lent him Immanuel Kant's *Critique of Pure Reason*, a book which generations of students of philosophy have found rather heavy going but which the teenage Einstein took to with ease.

Philosophy was to remain a lifelong interest for Einstein. His own adult writings are full of philosophical references and it is probably fair to say that he saw himself as being as much a philosopher as a scientist. Kant was only the beginning, and although Einstein remained deeply interested in the philosopher's work, he was to spend his late teens and early twenties immersing himself in the great philosophical tradition as a counterpoint to his purely scientific studies.

Einstein had become interested in what might be loosely

described as spiritual matters for at least a year before his introduction to Kant.

During this year he had undergone a religious phase. In addition to the Catholic education he had received at elementary school, he had been obliged by law to have private instruction in Judaism at home. However, the Einstein family did not observe many of the customs associated with their faith. For example, Hermann regarded the kosher dietary laws as anachronistic, superstitious nonsense. So it is highly unlikely that the twelve-year-old Albert was encouraged in his religious fervour by his parents, when he refused to eat non-kosher food and read the Bible avidly. They probably saw it as simply an adolescent fad which would soon pass. Many years later, when Einstein was a professor in Berlin, he told a friend that during this religious time he had composed a number of songs in celebration of his faith and sung them to himself on the way to school each day.

Einstein's religious fervour ended abruptly. Awakening common sense made him realise that much of what he had blindly accepted in religious books clashed with his growing scientific awareness. From about the age of thirteen, his attitude towards organised religion turned into resentment. In his words, youth was intentionally being deceived by the state through lies. From this point onwards, Einstein the scientist began to emerge. We begin to see the freethinking, intellectual rebel who mistrusted all handed-down knowledge. The adolescent Einstein was coming to the conclusion that in the future he would need to stand outside the conventional pattern of things and try to discover from without exactly how the world worked. He was becoming aware that his destiny lay in the domain of the intellect and that his external, everyday life should remain as uncluttered as possible. As he himself put it: 'Perception of this world, by thought, leaving out everything subjective, became partly consciously, partly unconsciously, my supreme aim.'[6]

Much of the science that Einstein was taught during the next few years would be familiar to children of modern generations. It involved the study of forces and motion based on the work of Isaac Newton, two centuries before. Because of the way the subject was taught, Einstein found it as tedious as many students still do (for the same reason) today. Yet Einstein would soon go beyond Newton to establish a new paradigm for the subject. In order to appreciate the nature of the revolution that Einstein's work engendered, we first need to remind ourselves what science was like before he came on the scene.

CHAPTER TWO

Physics Before Einstein

Modern science is often said to have begun with the work of Isaac Newton, in the seventeenth century. Indeed, the kind of physics we first learn in school, and which apply in everyday life for planning the construction of a bridge over a river, or calculating the trajectory of a spacecraft flying from the Earth to the Moon, is called Newtonian physics. Newton laid down three basic laws which determine the way objects move, and formulated a law of gravity which explains how objects attract one another at a distance. He even established a principle of relativity, more than two hundred years before Albert Einstein came up with his first theory of relativity. But Newton also built on the work of predecessors, in particular the work of Galileo Galilei, who died in the year that Newton was born – at least, according to the calendar Newton's mother would have used.

The qualification is necessary because Protestant England had not yet, in 1642, changed from the old Julian calendar to the Gregorian calendar used then on the continent of Catholic Europe, and now in use worldwide. The old calendar had gradually got out of step with the march of the seasons, because there is not an exact number of days in one year (one year being the time it takes the Earth to complete one orbit around the Sun). The Gregorian calendar, with its cunning use of leap years, keeps much more closely in step with the natural 'clock' of the Earth's orbit. When the calendar was reformed, in order to start the 'new' calendar off correctly the date had to be changed, overnight, not by one day but by several. As a result, when Newton was born it was Christmas Day 1642 in England – the same year that Galileo died.[1] But it was 5 January 1643 on the continent, and according to the calendar which we now use.

Newton's world

Whatever the calendrical complications, the timing of Newton's birth gave his undoubted genius the opportunity to flourish. By the second half of the seventeenth century, new ideas, replacing beliefs that had held sway since the time of the ancient Greeks, were in the air. Before Galileo came on the scene, for example, everyone believed that a heavier object would fall faster than a lighter object. There is no evidence at all to support the folk legend that Galileo disproved this by dropping a cannonball and a musket ball from the Leaning Tower of Pisa. But he did produce an elegant argument to show that all objects must fall at the same rate, if we ignore air resistance.

If you drop a large stone, it falls as one lump, at a certain rate. Now, if you break the stone in two equal parts, then hold the two halves together along the join, before dropping them together, what will happen? According to the old argument of Aristotle, each of the two halves will fall at the same rate, because they have the same weight. But because each is lighter than the original whole stone, they will fall more slowly than the original stone. But the only difference between the two halves and the original stone is that the original has been broken and then put back together. So either the reconstructed stone falls more slowly than the same stone does when it is in one piece, or Aristotle was wrong and all falling objects fall at the same rate.

Galileo actually studied the way things fall not by dropping them off the Leaning Tower, but by rolling balls down ramps. This slowed down the rate at which the ball descended, and gave Galileo time to measure how its speed changed, using the simple water clocks that were the best timepieces he had. He found that if you let a ball roll down a steep ramp and then up a shallow ramp, it would roll back up to the same height above the ground that it started from. This is exactly like the swing of a pendulum, reaching the same height on each side of the swing. But Galileo realised that if you make the shallow side of the ramp shallower and shallower, and ignore the effect of friction, then the rolling ball has to roll further and further along the ramp before it gets back up to the height it started from. Indeed, if the ramp is so shallow that it is actually flat, then the ball will roll for ever.

Galileo inferred that the natural motion of a freely moving object is to continue in a straight line until something stops it. This overturned the Greek idea that the natural state of any object is to sit at rest, stationary, unless something sets it in motion.

Galileo used these discoveries to explain the flight of a cannonball through the air. Before his time, it had been widely believed that a cannonball fired from a gun travelled horizontally through the air until its speed was 'used up', then fell vertically to the ground below. But Galileo said that a cannonball would follow a curved path – a trajectory – made by combining two motions. The horizontal motion would continue in a straight line, if it could (until the cannonball hit something), but there would be a vertical motion as well, making the ball drop like a stone from the moment it left the mouth of the cannon. The combination of these two motions would make the ball follow a curved trajectory (a parabola) until it hit the ground. And it was the impact with the ground that stopped the horizontal component of the motion.

Newton extended this idea with a powerful piece of imagery. He imagined a cannon mounted on top of a tall mountain, and so powerful that it could fire the ball as hard as you liked. He realised that, because the Earth is round, it would be theoretically possible to fire a cannonball so hard that the curve of its falling trajectory would exactly match the curve of the Earth's surface, falling away beneath it. The cannonball would circle the Earth, and arrive back at its starting place. This is exactly the way in which an artificial satellite, or the Moon itself, orbits around the Earth, constantly falling in a curve but never hitting the ground below. And Newton explained the nature of the Moon's orbit, and the fall of a cannonball (or an apple) with his theory of gravity, which jumped off from the work of another predecessor, Johannes Kepler.

Kepler had discovered the true nature of the orbits of the planets around the Sun, using painstakingly compiled observations recorded by the Dane Tycho Brahe. By this time, early in the seventeenth century, it was widely accepted (even if not formally acknowledged by the Catholic Church) that the planets moved in orbits around the Sun. Astronomers guessed that these orbits must be circular, with the Sun at the centre. But Kepler's careful study of the orbits, starting with Mars, showed that this is not so. In fact, the orbits of the planets are ellipses.

An ellipse is the curve you draw if you loop a piece of string around two drawing pins stuck in a board, and then use a pencil to pull the string into a tight triangle shape, and trace out a curve around the two drawing pins with the string kept tight. Instead of having one centre, like a circle, an ellipse has two foci (at the positions of the drawing pins, in this case). In the case of planets orbiting the Sun, the Sun is at one focus of the ellipse, while the other focus is empty. Kepler also found that the speed of a planet in its orbit varies in a regular way, so that it moves faster when it is closer to the Sun. And he found a third law relating the time it takes each planet to go around the Sun once (its 'year') to its average distance from the Sun.

Using Kepler's orbital calculations, Newton showed that the motion of the planets could be explained if, first, they had a natural tendency to move in straight lines at constant speed (like Galileo's rolling balls), and, secondly, they were being attracted by the Sun by a force proportional to one over the square of the distance between the planet and the Sun at any time. This is the famous inverse square law of gravity; it means that, for example, if the planet is magically moved to an orbit twice as far from the Sun, the force of attraction is not halved but reduced to a quarter of its previous value, since 2^2 is 4. And Newton showed that this same force of gravity, obeying the same inverse square law, could not only explain the motion of the Moon about the Earth, but also the fall of an apple from a tree. It is a *universal* law, applying to everything.

Newton also thought that there must be universal standards

of space and time against which motion, including the motion of the planets in their orbits, could be measured. His famous three laws of motion state that every object stays at rest or in a state of uniform motion in a straight line unless a force acts upon it; that if a force does act on an object, then the object accelerates (acceleration means a change either in the direction of motion, or the speed, or both) at a rate given by dividing the force by the mass of the object; and that if one object (physicists often use the term 'body', even when referring to inanimate objects) exerts a force on another body, then the second body exerts an equal and opposite force on the first. So the force of gravity of the Earth pulling on the Moon, for example, is balanced by the force of gravity of the Moon pulling on the Earth, while the deviation of the Moon's trajectory from a straight line is explained by the action of that force.

Newton envisaged that all this activity took place against a background of 'absolute space', which was the same everywhere and always at rest. The Earth and the Moon move through this absolute space, in the Newtonian world, and their speed relative to absolute space is measured in terms of absolute time, the imagined ticking of some ultimate cosmic clock. This is the bedrock of Newtonian relativity, the concept (to Newton) of literally God-given absolute standards of space and time against which everything else could be compared.

But this package of ideas – three laws of motion and a theory of gravity set in a framework of absolute space and time – which was to be the foundation of physical science for more than two centuries, was, in a sense, incomplete. If you throw a stone, you can tell that the force that set the stone in motion came from the action of your muscles. If a cannonball is fired from a gun, the force clearly comes from the explosion and the hot gases that push the ball out of the barrel. But how does the force of gravity reach out across empty space from the Sun to hold the planets in their orbits? Newton didn't know, and he wasn't worried about his lack of knowledge. He commented: '*hypotheses non fingo*', meaning 'I do not frame hypotheses', about the way in which gravity worked. He was satisfied that the inverse square law of gravity did provide a description of the way planets and moons are held in their orbits, without worrying about the way in which it did the trick.

This is where Einstein went further than Newton. But before we get on to Einstein's contributions to science, there are two more pillars of pre-Einsteinian physics that we have to set in place. The first of these is the nineteenth-century understanding of the nature of light.

Maxwell's waves

In Newton's day, light was thought to consist of a stream of tiny particles, which radiated outwards from a source of light and bounced off other objects to illuminate them. His contemporary, the Dutch physicist Christiaan Huygens, did argue that light might be explained as a kind of wave, spreading out from a source like ripples from a stone dropped in a pond. But Newton himself explained all the properties of light – the way that it travels in straight lines and how it reflects from a mirror, for example – in terms of the behaviour of streams of tiny particles. Throughout the eighteenth century, the particle theory of light held sway.

Early in the nineteenth century, however, the English physicist Thomas Young, quickly followed by the Frenchman Augustin Fresnel, carried out experiments which seemed to prove conclusively that light must, after all, be a form of wave. The classic example of this proof is known as the 'Young's double-slit experiment', repeated today in every school physics laboratory. This is the optical equivalent of allowing waves on a pond to pass through two small holes in a barrier. The waves will spread out in a series of overlapping semicircles from each of the two holes, and the overlapping sets of waves will interfere with each other to produce a complicated new pattern. In some places, the waves cancel each other out, and the surface of the water is undisturbed. In other places, the two waves add together, and the surface of the water is agitated more than it would be by either wave alone. When Young allowed a beam of pure light to pass through two tiny slits in a sheet of card, he found that the light coming through the two slits formed a pattern of bright and dark stripes on a screen. This pattern can be precisely explained in terms of the interference of light waves spreading from the two slits, just like the interference of ripples on a pond. There is no way that this interference pattern could be produced by tiny particles of light like those envisaged by Newton.

It took so long to prove that light is a wave because the wavelength of light is very small. The interference effects are visible only if the slits the light passes through are as narrow as the wavelength of light – and the wavelength of ordinary visible light is measured in tens of *millionths* of a metre. Once this was appreciated, it turned out that all the properties of light which Newton had explained in terms of particles could be explained just as well in terms of waves.

Even in Newton's day, however, physicists had a good idea of how fast light moved, whatever it consisted of. The measurements stemmed from another of Galileo's discoveries, the existence of moons orbiting the planet Jupiter. But the measurement was not made by Galileo himself, but by the Dane Ole Rømer.

It all depends on the fact that from time to time a moon of Jupiter will go behind the planet, and be eclipsed, as seen from Earth. Even before astronomers were armed with Newton's theory of gravity, they could observe these eclipses and, because the eclipses follow a regular pattern, work out a timetable predicting when the next eclipse was due. They were puzzled to find that this timetable was almost, but not quite, regular, and that there were discrepancies. Sometimes the eclipses seemed to occur a few minutes late, sometimes they were a few minutes early. Rømer explained this in terms of the Earth's motion around the Sun, and the speed of light.

When the Earth is on the same side of the Sun as Jupiter, light from the moon that is being eclipsed does not have to travel so far to reach the Earth as it does when the Earth is on the opposite side of the Sun. The extra distance is the diameter of the Earth's orbit. So the extra time taken for the 'news' of the eclipse to reach astronomers on Earth is just the diameter of the Earth's orbit divided by the speed of light. Using the best estimates he had of the size of the Earth's orbit, Rømer calculated, in the 1670s, that the speed of light must be (in modern units) 210,000 km per second; using modern estimates of the size of the Earth's orbit around the Sun, the figure is 300,000 km per second. And the speed of light has now been measured many times using other techniques, always giving the same figure (which means that we can now measure the size of the Earth's orbit from the timing of those eclipses of the moons of Jupiter, turning Rømer's technique on its head).

So, by the 1820s physicists knew that light consisted of a wave moving at a speed of 300,000 km per second. But what, exactly, was waving? And what was the wave moving through? Where, if you like, was the pond on which the ripples of light moved? The answer came from an unexpected source.

Before the nineteenth century, physicists had been intrigued and puzzled by the seemingly similar phenomena of electricity and magnetism. Electric charges come in two varieties, dubbed 'positive' and 'negative', and the same kind of charges (two positives or two negatives) repel each other, while the two opposite kinds attract each other. Magnetism also comes in two varieties, dubbed 'north' and 'south', although, unlike electric charges, the two varieties cannot exist independently and every north magnetic pole has a south pole associated with it, making a 'dipole', like an ordinary bar magnet. As with electricity, though, the same kinds (two north poles or two south poles) repel each other, while opposite poles attract. And in both cases, as with gravity, the force involved varies inversely as the square of the distance between the two charges, or two magnetic poles. But in spite of these similarities, a magnet placed near an electrically charged object seems to have no effect on it, and vice versa.

It was only in 1820 that the Danish physicist Hans Christian Ørsted found the first connection between electricity and magnetism. He discovered that when an electric current – a flow of electric charges – is moving along a wire, there is a magnetic field wrapped around the wire. This will disturb a magnet (such as a compass needle) placed nearby. In other words, a *moving* electric charge creates a *static* magnetic field.

In 1831, the English pioneer Michael Faraday (quickly followed by the American Joseph Henry and then by the Russian

H. F. E. Lenz) discovered the opposite effect, that a magnet moved past a wire will induce an electric current in the wire. In other words, a moving magnet creates an electric field, and the charges in the wire (now known as electrons) move in response to that field.

In the same year that Faraday made this discovery, the Scot James Clerk Maxwell was born. Three decades later, in the 1860s, Maxwell used these discoveries to help him construct a mathematical theory – a set of equations – that explained both electricity and magnetism in one package, electromagnetism. This was the first example in physics of the successful unification of two forces of nature into one mathematical description. Maxwell's equations described electricity and magnetism in essentially the same way – as far as the two forces are concerned, the equations are symmetric. Maxwell realised that this led to a completely unexpected implication of the theory. In effect, a varying electric field could give rise to a varying magnetic field, which would in turn give rise to a varying electric field, and so on. Once the varying electromagnetic field had been created by an input of energy, it would propagate, spreading out as an electromagnetic wave.

How fast would it propagate? The solution to Maxwell's equations included a constant, which was the speed with which those electromagnetic waves must travel. Using the measured properties of electric and magnetic fields, Maxwell could calculate the value of that constant. It turned out to be the speed of light, 300,000 km per second. There could be no doubt that light itself must be a form of electromagnetic wave. In unifying electricity and magnetism, Maxwell had explained the nature of light. What is 'waving' in a beam of light is a linked pair of varying electric and magnetic fields.

At that time, in the latter part of the nineteenth century, physicists thought that the universe – empty space itself – must be filled with a mysterious substance, called the ether, in which the electromagnetic waves rippled. The ether was thought to be rather like Newton's absolute space – a reference frame against which motion could be measured. The speed of light that came out of Maxwell's equations ought, then, to be the speed of light through the ether. And since the Earth moves in a closed orbit around the Sun, it must be travelling through the ether at different velocities at different times. So, for example, when the Earth is in such a position in its orbit that it is moving towards Jupiter, light signals bringing news of an eclipse should be travelling, relative to the Earth, a little faster than their speed through the ether, and will arrive a little early (over and above the effect calculated by Rømer). Similarly, when the Earth is moving away from Jupiter, the light signals, struggling to overtake the receding Earth, should arrive a little late.

Maxwell died in 1879, the year Einstein was born, and never saw experiments along these lines carried out. But by the end of the nineteenth century, experiments involving observations of subtle changes in interference patterns like those produced in Young's slit experiment had become sophisticated enough to measure the tiny differences that ought to have been produced in this way by the Earth's orbital motion. The snag was, none of these experiments showed any such effect, and none ever has. As far as anyone can tell, the speed of light is always exactly the same, as measured on Earth, whichever way the Earth is moving 'through the ether', and whether the light beam is overtaking the Earth or running into the Earth head on in its orbit.

So physics at the end of the nineteenth century was based upon two beautiful theories. The first, Newtonian mechanics, described the way physical bodies interact, and rested in part on the assumption that there was an absolute space which things moved relative to. The second, Maxwell's electromagnetism, explained the nature of light and other radiation (radio waves had been discovered by then, and explained as another form of electromagnetic wave, also propagating at the speed of light but with longer wavelengths than visible light), but had led to experiments which showed no trace of Newton's absolute space. Both theories worked fine in their place, but they did not agree with one another in the one area where they overlapped.

The reason for the discrepancy would soon emerge, and would result in a completely new understanding of the universe. But in the 1890s, thanks to those two great theories and the third pillar of physics, it had seemed, briefly, as if the scientific understanding of the basic laws of nature was essentially complete, and all that remained for future generations was to dot a few 'i's and cross a few 't's. That third pillar of nineteenth-century physics was, at the time, the most practical of all sciences – thermodynamics, the physics behind, among other things, the workings of steam engines.

Kelvin's contribution

Thermodynamics is different from the other two pillars of nineteenth-century physics because, instead of dealing with one or two bodies (a planet orbiting the Sun, or a couple of moving electric charges), it deals with the statistical behaviour of vast numbers of particles in realistic, everyday situations, such as those steam engines (an everyday phenomenon in the late Victorian era) or the way a cup of coffee cools down.

Today, we are used to the idea of everyday things, like a cup of coffee, being made up of tiny particles called atoms and molecules. An atom is the smallest particle of an element, such as hydrogen or oxygen, that can exist; a molecule is the smallest unit of a compound, such as the water (hydrogen oxide, or H_20) that makes up the bulk of your cup of coffee. So it is worth recalling that in the nineteenth century nobody could prove that atoms really existed, and there was a sometimes fierce debate on the subject. The arguments in favour of the existence of atoms were, at that time, statistical. Maxwell, and the Austrian Ludwig Boltzmann, explained the behaviour of gases in detail, using the concept of atoms and the pure application of Newton's laws of motion.

They showed how details of the way gases behave – not just the fact that a gas can be compressed, for example, but the precise amount by which it will be compressed if a certain pressure is applied – can be explained if the gas is made up of very many tiny, hard spheres (the atoms and molecules) that bounce around, colliding with one another and with the walls of any container that they happen to be in. Although any one molecule might be moving in any direction at any one time, the combined effect of a myriad of these tiny spheres bouncing off the walls every second would produce a steady pressure. This whole package became known as the 'kinetic theory' of gases, and is part of the science of thermodynamics.

Thermodynamics, as the name implies, relates heat ('thermo') to movement ('dynamics'). When gas in a piston is heated, for example, it will expand and push the piston outward, the basis of the steam engine that drove the Industrial Revolution. It converts heat into useful work. Among other things, thermodynamics also defines the flow of time in the universe, and suggests that in some sense the death of the universe itself may be inevitable.

Much of the pioneering work in thermodynamics was carried out in the nineteenth century by the British physicist William Thomson, who later became Lord Kelvin, and in whose honour the absolute scale of temperature used by physicists, the Kelvin scale, is named. As far as our deep understanding of the universe is concerned, the greatest discovery in thermodynamics is the so-called 'second law', which says that heat always flows from a hotter object to a cooler object when the two objects are placed in contact. It sounds blindingly obvious; yet this simple rule has been described as the supreme law of nature. It tells us which way time flows, and that things wear out.[2]

There are two classic examples which highlight the importance of this law. First, imagine a cup of coffee, into which you drop an ice cube. The ice cube melts, as heat flows from the hot coffee into the cold ice. You never, ever, see a cool cup of coffee start to get hotter, all by itself, while an ice cube grows in the middle of the cup – even though the cup of hot coffee plus the ice cube has exactly the same amount of heat energy, overall, as the cup of cool coffee with no ice cube. This is what we mean by saying that the second law defines the arrow of time. If you made a film of the melting ice cube, and ran it backwards through a projector, everybody who watched the film would know it was running backwards.

This is also an example of things wearing out. A cup of coffee with an ice cube in it has more structure than a cup of cold coffee, just as a house has more structure than a pile of bricks. Like the melting ice cube, if you leave a house unattended for long enough it will turn into a heap of bricks. But no matter how long you leave a heap of bricks unattended, it will never form itself into a house. This kind of structure possesses a quantity of information. The lost structure is like lost information, because the cold coffee has no 'memory' of the fact that some of the liquid used to be an ice cube, just as a brick has no 'memory' that it used to be part of a wall. Being slightly perverse, physicists actually measure information in a negative sense, and give negative information the name 'entropy'. The second law of thermodynamics can be expressed most succinctly as saying that entropy always increases – things wear out, ice cubes melt, houses fall down, people die.

In everyday life, we seem to be able to break this law. Indeed, life itself seems to break this law. We are born, grow and learn things, build houses and make ice cubes. But all this activity on Earth depends on one thing – a flow of energy from the Sun, heat flowing from a hotter place to a colder place, in accordance with the second law. The processes which release that energy inside the Sun also produce a vast amount of entropy, which more than offsets the tiny (on the cosmic scale), temporary reversal of the natural flow of entropy here on Earth. Taking the universe as a whole – indeed, just taking our Solar System as a whole – entropy *is* always increasing. Things are wearing out. One day, in the far distant future, when all the stars in the universe have burned out, life will be impossible, and the universe will suffer what is known as the 'heat death', when the temperature everywhere is the same and thermodynamic processes, which depend on heat moving from a hotter place to a colder place, will no longer take place.

But there is one strange feature of all this which we ought to mention. The second classic example of the second law at work is based on imagining a box of gas with a partition in the middle. One half of the box can be pumped out, so that it is empty, with no atoms in it. The other half is full of atoms, tiny little spheres moving rapidly and bouncing off one another in accordance with Newton's laws of motion. If we pull the partition out of the way, the atoms will spread out to fill up the entire box, and the temperature of the gas will fall as they do so. But if you start with a box full of gas, you will never see all the atoms move together up into one end of the box, leaving a vacuum in the other half.

Or will you? Boltzmann calculated that if the atoms are moving about at random there is, in fact, a very tiny chance that all the atoms in the box will happen to move together in this way. The chance of this happening is so small that it would take billions upon billions of years for it to happen. And this gives us another perspective on the universe. Low entropy states are possible, but they are extremely unlikely. On the other hand, the universe at large must be in a low entropy state, or the Sun and stars would not be radiating energy, increasing entropy, and providing the opportunity for life to exist on Earth. This is the other side of the heat-death coin. If the universe is not yet in a state of heat death, or thermodynamic equilibrium, then there must have been a 'heat birth' of some kind, when the universe came into being in a state of low entropy.

Boltzmann himself considered the possibility that the entire universe that we see around us may be no more than a random statistical fluctuation, a temporary bubble within an even vaster sea of heat death, where everything is in thermal equilibrium with maximum entropy. Like the atoms of gas all moving together into one end of the box, such a universe could conceivably arise as a result of an extremely rare random fluctuation, a bubble of space which is now slowly returning to its normal long-term state.

These prescient ideas were far ahead of their time. They suggested that the universe as we know it (our 'bubble') had both a finite past and a finite future, that it was born and that it must die. At the end of the nineteenth century, such ideas were dismissed by astronomers. It was generally accepted by them that the universe was essentially eternal and unchanging – individual stars might go through their life cycles and die, like individual trees in a forest, but overall the universe would always stay much the same, like the forest itself. Those ideas, like the false confidence in the near-completeness of the physical description of the world provided by Newton's laws and Maxwell's equations, were to be overturned within a generation. But in the mid-1890s, when this comfortable complacency was firmly established, the person who was to do the overturning was still in high school, and about to have the comfortable familiarity of his own world turned upside down.

In 1894, when Einstein was fifteen, his father decided that the family business should move to Milan, and that Albert would be left behind in Munich to complete his education at the Luitpold Gymnasium.

Chapter Three

College Drop-Out

The idea for the business move initially came from the Italian representative of the Einsteins, Signor Garrone, who suggested that prospects would be better in Italy. Hermann Einstein was unconvinced at the outset but was swayed by the enthusiasm of his brother, and it was decided, in June 1894, that the house in Munich should be sold and the entire family (with the exception of Albert) moved south.

Because Albert was now at a crucial stage in his education, with matriculation in sight, Hermann concluded that his son should not be uprooted by joining the family until he had completed his final three years at the gymnasium. Albert's Italian was minimal and the suggestion that he might attend an Italian school was deemed impractical. Instead he was installed in a boarding house in Munich and watched over by a relative.

Hermann and Jakob meanwhile began to set up a factory in Pavia under the new business name of 'Einstein and Garrone'. Once again, much of the finance came from the extended family and optimism outweighed many practical considerations.

Albert was very unhappy about being left in Germany. He hated the gymnasium and had loathed the prospect of another three years there even when he had the comfort of a close-knit family environment. Now, with what must have seemed an eternity of rote learning, severe discipline and tedium stretching ahead of him, he appears to have fallen into a deep depression.

He lasted in Munich for only six months and his method of escape was typically ingenious. After a particularly irksome period at school, he managed to persuade a doctor to supply him with a medical certificate which stated that he was suffering from a nervous disorder. This provided him with a valid excuse for leaving the gymnasium, so that it would not appear that he had merely run away from the situation in which he found himself. However, when Einstein announced to the principal of the gymnasium that he was leaving, he was curtly informed that he was in fact expelled.

Having the rug pulled out from under his feet in this way incensed the teenager, and many commentators have suggested that it was this humiliation that for ever turned him against the German way of life.

The reason for Einstein's expulsion appears in retrospect to be somewhat flimsy – that his presence in class was disruptive and affected the other students. One of his teachers had declared in front of the other students that he would have preferred it if Einstein had not been in the class. When Einstein then asked why, since he had done nothing wrong, the teacher replied: 'Yes, that is true. But you sit in the back row and smile, and that violates the feeling of respect which a teacher needs from a class.'

There was another reason for his desire to leave Germany. German law stated that if a boy left the country before the age of seventeen he would be exempt from national service. Einstein detested regimentation in thought or in action. This had first shown itself when his father had taken him, as a very young boy, to see a military parade. The automatonlike movements of the soldiers marching in strict formation disturbed the boy so much that he burst into tears and had to be taken home.

The idea that he might now be subjected to military service was complete anathema to him. Combined with his distaste for the teaching methods of the gymnasium and the fact that he missed his parents and sister Maja, this prompted his decision to leave.

Unannounced, he turned up in Pavia in the early spring, shortly after the family had moved there from their original residence in Milan. Not surprisingly they were shocked and dismayed at the fact that Albert had taken it upon himself to behave in such a high-handed manner. Barely sixteen, he had gone completely against his family's plans for him. And there were more shocks in store for Einstein's parents. Within days of arriving in Pavia, he told his father that he had decided to renounce both his German citizenship and his Jewish faith.

Both these moves were of course part and parcel of his avowed intention to maintain an isolation from the world and to be the master of his own destiny.

Some have seen this as near-paranoia, as 'an emotional fissure which split Einstein's character from end to end like a geological fault'.[1] The psychologist Anthony Storr sees the action as another symptom of Einstein's schizophrenia, and more than one commentator has made the point that his gentle character was apparently at odds with his loathing for the Prussian Empire, his country of birth, and all that it stood for.

Renouncing his German citizenship was of paramount importance to Einstein. Hermann Einstein was reluctant to give in to his son on this matter but gradually accepted that Albert would go through with it without his help when he had come of age. After a great deal of persuasion Albert did gain the agreement of his parents, although Hermann Einstein did not send off the required papers until Albert was seventeen and at school in Switzerland. Consequently, from the age of seventeen until he was later naturalised in Switzerland, he was a stateless person. This fact is contested by several authorities, most especially André Mercier, head of the department of theoretical physics at the University of Bern and secretary-general of the International Committee on General Relativity, who has stated that, by German law, a minor could not renounce his nationality and that Einstein was therefore a German citizen until he had come of age.

As a measure of the family's unorthodox approach to their inherited faith, Einstein's parents were almost totally unconcerned that Albert had also decided to reject Judaism.

Faith and institutionalised religion were completely at odds with the two prime motivating forces in Einstein's personality – his scientific understanding and his abhorrence of authority. In later life, as a Zionist sympathiser, he was a supporter of the Jews as a race rather than as a religious movement.

Albert Einstein as a teenager had a distinct arrogant streak.

Fortunately, in Pauline and Hermann Einstein, he had unusually liberal-minded and intelligent parents who were quick to realise that the best way to treat their son was to allow him to find his own way and certainly not to be heavy-handed with him. As part of a close-knit family, they understood his character and his need for freedom of thought and expression. Yet they were quite naturally concerned that the boy was going to fritter his life away. Here he was, just sixteen. He had left school without a qualification, his head was full of what Hermann Einstein saw as 'philosophical nonsense', he was determined to renounce his nationality and was drifting further from parental control every day.

There is little doubt that Einstein's untimely expulsion from the gymnasium simultaneous with the triumphant announcement of his own rejection of the place fanned the flames of his hatred of Prussia. This may itself be viewed by some as a little reactionary, but then, Einstein always had been a sensitive soul and, coming at a time when he was entering the arrogance of his teenage years, it may not be too surprising.

Einstein succeeded in quelling his parent's fears for his future by reassuring them that he had no intention of deserting education and that, in fact, by leaving the gymnasium he was creating the best environment for progressing with his programme of self-education. To prove his sincerity in the matter, he announced his intention to take the entrance examination at the Federal Institute of Technology (the ETH – Eidgenössische Technische Hochschule) in Zurich the following autumn.

In the meantime Einstein gained his parents' permission to develop his cultural education by embarking on a solo trip around the main art centres of Italy. He travelled to Pisa and Siena and wrote home declaring his enthusiasm for works of art he saw there, especially the paintings of Michelangelo. He also enjoyed the company of the Italian people, their approach to life and their relaxed manner, so at variance with his Teutonic upbringing.

During his travels, Einstein did not forget his scientific ideas. He wrote a lengthy letter to his uncle Cäsar, outlining his ideas about many of the areas of physics that fascinated him, including the relationship between electricity, magnetism and, most significantly, the concept of the ether, which would occupy much of his intellectual efforts throughout his time at college.

Upon his return to his parents, he began his preparations for the ETH entrance examination, and in the autumn of 1895 he set off for Zurich.

There are conflicting accounts of Einstein's first visit to the ETH. Suffice it to say that he failed to gain entry to the college. This was perhaps not particularly surprising. He was, after all, taking the exam a full eighteen months ahead of other candidates. The exam covered a wide range of subjects and, although he was noted as having done exceptionally well in mathematics and sciences, he did not reach a sufficiently high standard in languages, history, literature and art to qualify for the course.

It has been suggested that he deliberately failed the entrance exam, knowing that his father was keen for him to begin a course in engineering if he had been successful, and that he needed to buy time to convince his family that he should be allowed to enter a pure science or mathematics course. In any case, the principal of the ETH, Albin Herzog, had been impressed by Einstein's obvious scientific and mathematical abilities and recommended to his parents that Albert spend a year in a Swiss secondary school and then retake the examination.

The following year was one of the happiest of Einstein's youth. The school chosen by his family, upon the advice of an old friend of Hermann Einstein's, was the cantonal school in the small Swiss town of Aarau, some twenty miles to the west of Zurich. For the first time in his education, Einstein had found a school that perfectly suited his temperament. It was a small country school which operated in a relaxed and friendly atmosphere. The principal was Professor Jost Winteler, a liberal-minded man and a highly respected teacher who treated his pupils as adults and approached education with a free thinking manner which Einstein immediately found endearing.

For the duration of his stay in Aarau, Einstein lodged in the home of Professor Winteler and his family. A close and lasting relationship developed between Einstein and the Wintelers and he addressed Jost and Pauline Winteler as 'Papa' and 'Mutti'. The friendship extended to the rest of the Einstein family and fourteen years later, in 1910, Einstein's sister Maja married Professor Jost Winteler's son Paul.

In January 1896, Einstein received official acceptance (upon payment of three marks) of his renunciation of German citizenship, and for the next five years he was a stateless person. Einstein was delighted.

The image of Albert Einstein handed down from 1896 in Aarau is of a self-assured, almost overconfident young man. He was considered to be handsome with fashionable looks, although even at that age he paid little attention to the way he dressed. He was very popular at school and liked by the other pupils and his teachers. Perhaps an element in his air of confidence during this time was the knowledge that he had finally escaped the repression he had felt in his homeland. He was developing a sharp tongue and a line in snappy answers to questions he viewed as unworthy of a civilised reply. This trait appears to have been tempered somewhere along the line and grew into an easy wit and verbal deftness later in life.

Certainly the physical environment could only have been beneficial to his sense of wellbeing. Along with his classmates, he would regularly go on long walks in the mountains. The air was clean and the scenery breathtaking. For the first time in his life he felt relaxed about learning outside of his own private intellectual journeys. In old age he was to remember his time at school in Aarau with great fondness.

His formal work showed signs of application, and his plans for the future were as steadfast as they had always been. In a short French essay which had been set for homework, '*Mes Projets d'avenir*', he succinctly outlined his game plan.

> My plans for the future
>
> A happy man is too content with the present to think much about the future. Young people, on the other hand, like to occupy themselves with bold plans. Furthermore, it is natural for a serious young man to gain as precise an idea as possible about his desired aims.
>
> If I were to have the good fortune to pass my examinations, I would go to [the ETH in] Zurich. I would stay

> there for four years in order to study mathematics and physics. I imagine myself becoming a teacher in those branches of the natural sciences, choosing the theoretical part of them.
>
> Here are the reasons which led me to this plan. Above all it is [my] disposition for abstract and mathematical thought, [my] lack of imagination and practical ability. My desires have also inspired in me the same resolve. That is quite natural; one always likes to do the things for which one has ability. Then there is also a certain independence in the scientific profession which I like a great deal.[2]

Einstein's enthusiasm and application at the little school in Aarau paid off. At the end of the year, in the early autumn of 1896, he obtained his diploma enabling him to enrol on a course at the ETH. He received top marks (6 out of 6) for history, geometry, descriptive geometry and physics; 5 for German, Italian, chemistry and natural history, and 4 for geography, art and technical drawing.

Rather than enrolling on a pure-science course, and perhaps also in order to reach a compromise with his family, Einstein decided on a teaching course which would, after four years, qualify him to be a specialist teacher in physics and mathematics at secondary level. Despite his intellectual manifesto and desire to reach the heights of understanding in science, his worldly ambition was surprisingly modest. At the age of seventeen he had little notion of where his journey would take him other than the vague idea that he would like to pass on his scientific understanding to others.

As now, Zurich in 1896 was the commercially most important city in Switzerland; the centre of the banking trade and the country's largest city. Built on both banks of the river Limmat as it issues from the northern tip of Lake Zurich, it is known to have been settled before the Romans fortified the site around the first century AD. In 1896, the region which constituted the city proper had only recently been expanded in order to incorporate eleven outlying communities.

The ETH was a highly respected technical college with

a particularly well-organised and superbly equipped physics institute. The department had been established by the physicist Professor Heinrich Weber, who still taught there, and his friend, the engineering tycoon Werner von Siemens, who had died a few years before Einstein entered the college.

An American physics professor, Henry Crew, who visited the ETH a few years before Einstein joined, said of the place:

> H. F. Weber and Dr Pernet are at the head of the physics department in the Polytechnic. They not only have the most complete instrumental outfit I have ever seen, but also the largest building I have ever seen used for a physical laboratory. Tier on tier of storage cells, dozens and dozens of the most expensive tangent and high resistance galvanometers, reading telescopes of the largest and most expensive form by the dozen, 2 or 3 in each room. The apparatus cost 400,000 francs, the building alone 1 million francs.[3]

Einstein arrived in Zurich on 29 October 1896. For the first two years he stayed at Unionstrasse 4 in the house of a Frau Kagi. His final two years were spent first at the home of Frau Markwalder, followed by a return to Frau Kagi, who had by this time moved to a new address. Throughout his four years at the ETH Einstein lived on 100 Swiss francs per month. Most of this came from the Kochs. It was not a princely sum but adequate for his needs. Small furnished rooms in Zurich at the time cost about 20 francs per month, whereas a boarding house supplying two hot meals daily would cost in the region of 40 francs per month. What did cause him some difficulty was having to save 20 francs per month to pay for his Swiss naturalisation papers.

From all accounts Einstein was a typical student in the Zurich of the turn of the century. He dressed sloppily, tended to avoid as many lectures as possible and spent a great deal of his time in cafés and bars, deep in discussion with fellow students. The environment at the ETH was relaxed, but Einstein was expected to maintain a high academic standard and at least give the impression that he was behaving in a suitably studious fashion. This image also had to be presented to his family, who were, after all, paying for him to be there. Indeed, he did work hard during his time in Zurich, but worked as he had always done by going his own way. He was highly selective about which lectures he attended and for whom he completed work. Einstein said of his efforts at the ETH:

> There I had excellent teachers (for example, Hurwitz, Minkowski), so that I should have been able to obtain a mathematical training in depth. I worked most of the time in the physical laboratory, however, fascinated by the direct contact with experience. The balance of the time I used, in the main, in order to study at home the works of Kirchhoff, Helmholtz, Hertz, etc. The fact that I neglected mathematics to a certain extent had its cause not merely in my stronger interest in the natural sciences than in mathematics but also in the following peculiar experience. I saw that mathematics was split up into numerous specialities, each of which could easily absorb the short lifetime granted to us. Consequently, I saw myself in the position of Buridan's ass, which was unable to decide upon any particular bundle of hay. Presumably this was because my intuition was not strong enough in the field of mathematics to differentiate clearly the fundamentally important, that which is really basic, from the rest of the more or less dispensable erudition. Also, my interest in the study of nature was no doubt stronger; and it was not clear to me as a young student that access to a more profound knowledge of the basic principles of physics depends on the most intricate mathematical methods. This dawned upon me only gradually after years of independent scientific work.[4]

Einstein had a great deal of respect for many of his teachers at the ETH. He was particularly impressed by the teaching and the ideas of his mathematics lecturer, Hermann Minkowski. This man would a few years later be instrumental in helping to establish a strict formalism for the theory of relativity, but was once quoted as describing Einstein as 'a lazy dog who never bothered about mathematics at all'.

Einstein's admiration did not extend to the top man in the

physics department, Heinrich Weber. By all accounts the bad feeling was mutual. Weber once said to the eighteen-year-old Einstein: 'You're a smart boy, Einstein, a very smart boy. But you have one great fault: you never let yourself be told anything!' The trouble with Weber appears to have been rooted in the fact that he was old-fashioned in his approach and in his scientific interests. Weber was said to have simply ignored any physics after the work of Hermann von Helmholtz, who had made his great discoveries in electricity and magnetism in the 1850s. Weber also dampened Einstein's enthusiasm for experiment. Almost from the moment he arrived at the ETH he had the desire to conduct experiments on the Earth's movement against the ether. Yet, shortly afterward, Einstein received a strong warning from his teachers that he was ignoring his practical work. As at the gymnasium in Munich, Einstein turned against the subject in which he felt frustrated.

Despite the dissuasion of his teachers, Einstein did set up a laboratory experiment in the department in mid-July 1899, towards the end of his third year. However, he pushed the equipment beyond its limits and something gave. As a consequence he seriously injured his right hand, which needed several stitches at the city clinic. Einstein seems to have been a little embarrassed by the accident. He confided in a friend, Julia Niggli, with whom he frequently corresponded, but begged her not to pass on the news to the Wintelers, presumably because they would have become unnecessarily concerned for his future safety in the laboratory. This incident may also have put Einstein off experimental work.

Although he got on better with many of the other professors at the ETH, the antagonism between Einstein and Weber never really abated. Einstein cockily called Professor Weber 'Herr Weber' to his face, rather than the polite 'Herr Professor', knowing that Weber hated being addressed in this comparatively disrespectful manner.

When Weber died in 1912, Einstein wrote to a friend that 'Weber's death is good for the ETH'.[5]

During Einstein's first term at the ETH, the family business was once again in serious trouble. The factory in Pavia had to be closed and the business liquidated. This time the failure was almost total. Jakob decided that he had had enough and, after sixteen years in business with his brother, he decided to gain employment with a large engineering firm. Hermann, however, took the decision to carry on.

Albert tried to dissuade his father from this course of action and even went to visit relatives, who were considering financing Hermann's new project, to influence them against lending his father any more money. He was unsuccessful and Hermann moved the family to Milan to try once more. Almost predictably, this latest venture met with failure two years later. Hermann Einstein then managed to re-establish himself, this time helping to install electrical power stations.

These family difficulties haunted Einstein's early days at the ETH. He felt that he was leading a privileged life of higher education while his poor family were struggling to survive and moving from town to town. He felt that the only thing he could do was to apply himself totally to his own programme and to gain as much as he could from the limited approach of the college.

To many, Einstein's stance would appear to be extremely arrogant, his distrust of a regulated system neurotic and high-handed. However, it could also be said that he simply knew very well his own mind and had, at an early age, discovered what was right for him and how best he could learn and understand the way the world worked. He was arrogant and, as a teenager, more than a little full of himself, but he was a young man with a highly tuned and remarkably mature self-awareness, as his eventual great success proved.

Zurich was a stimulating place at the turn of the century, and a dynamic cultural centre. Many intellectuals from all spheres took up residence in the city, including Leon Trotsky, Rosa Luxemburg and, later, Lenin and James Joyce; others appeared fleetingly. Although his resources were limited, Einstein made the most of the environment in which he found himself. His musical ability was a help, serving as an introduction to a broader social circle. He also began to read a great deal. This was not merely to serve as a counterpoint to his scientific studies but grew out of a genuine interest in the arts, an interest which he maintained throughout his life. He was particularly fond of the novelists Balzac, Dickens and Dostoyevsky, especially *The Brothers Karamazov*.

He made a number of friends in Zurich, some of whom were greatly to affect his future. The three closest were Marcel Grossman, a dedicated and studious young man whose father owned a factory manufacturing agricultural equipment; an engineering student, Michelangelo Besso, whom Einstein met several months after his arrival in Zurich and with whom he enjoyed a lifelong friendship; and Mileva Maric, whom Einstein married some six years later. She was a quiet, serious student a few years older than Einstein, and had joined the college the same year as he.

In an era when it was still relatively rare for women to pursue any form of higher education, let alone a technical subject, Mileva Maric was an unusual person. In addition, she came from a relatively poor Swabian family living in southern Hungary. The fact that she had reached this stage in her education indicated exceptional reserves of determination and motivation as well as a capable mind.

Descriptions of Mileva vary widely, but the general consensus is that she was a retiring and often distrustful woman who was difficult to get to know but who showed hidden depths of character once won over. She was not unattractive, but neither was she any great beauty. She had dark eyes and a mass of thick dark hair which she usually wore in a tight bun. She had a limp, but this was only slight and has been exaggerated by unkind portraits of her.

With her solid, down-to-earth approach to life, Mileva was undoubtedly a stabilising influence on Einstein. She worked very hard at college and was determined to succeed after all the efforts it had taken her to get there. She was in fact quite at variance with the type of woman Einstein usually found attractive. He was very fond of female company, and was said to have felt more comfortable with women than with men – a characteristic which remained into old age. However, he was not usually very interested in intellectual women. Mileva appears to have been one of the rare exceptions, though she was not a particularly brilliant student. She twice failed to obtain her diploma, first in 1900 with the others of her group and then again in 1901.

If the existing letters between Einstein and Mileva before they were married are anything to go by, her personality was certainly not easy to access. They were frequently separated, visiting their respective families during vacations, and wrote a great many letters to one another. It appears that it was Einstein who was doing most of the running in their courtship. It is possible to see thinly veiled impatience at the fact that Mileva was often slow to respond to his letters and that she kept herself very much to herself; for example, in this opening to a letter from Einstein to Mileva written some two years into their relationship, in 1899:

> Milan, Tuesday 10 October 1899
>
> DSD [Dear Sweet Doxerl]
>
> Now, you are a fine one! It's already the 4th day that she has been sitting very cozily at the examination and has not yet uttered a single word to her good colleague and coffee-guzzling pal. Isn't that horrible? I shall compose a fire-and-brimstone sermon and give it to you in person next Monday, and early in the morning at that. And if the maid says that you have left and I see your polished little boots in front of the door, which seems to be happening from time to time – then I'll simply wait a little or I'll get a shave . . .[6]

('Dear Sweet Doxerl' has been suggested by the translator of the original letter. Doxerl was a nickname Einstein often used for Mileva.)

During the following year, some time between the autumn of 1899 and the summer of 1900, the relationship stepped up a gear. By this time, the couple were writing to each other in much more affectionate terms. Albert has mentioned their relationship to his parents, who are concerned for his future and the effects of a love affair on his college work.

There had been other women in Einstein's early years at the ETH; he was popular among the female element of the college and women he met through his male friends. These relationships could of course have been quite platonic, but this is open to conjecture. One of the women with whom he seems to have grown close was Julia Niggli, to whom Einstein gave advice when she became involved with a young man. In a letter to Miss Niggli, dated 28 July 1899, Einstein says at one point: 'You certainly seem to be in an enchanting little nest at present. I would love to help you kill time there in all sorts of pleasant ways.'[7]

Another woman with whom Einstein had a flirtation was Anna Schmid. In her album he wrote:

> You girl small and fine
> What should I inscribe for you here?
> I could think of many a thing
> Including also a kiss
> On the tiny little mouth.
>
> If you're angry about it
> Do not start to cry
> The best punishment is –
> To give me one too.
>
> This little greeting is
> In remembrance of your rascally little friend.
>
> Albert Einstein[8]

A bit of a rascal he certainly seems to have been during his college days. He made quite an impression on those around him. He was a large young man with strong features, a great shock of black hair and a raffish moustache. He was lively and witty, he had travelled and had experienced something of life during his journeys in Italy. As well as having a reputation as a rebellious but highly gifted student of science, he also played the fiddle and piano, loved to perform before his friends and enjoyed the company of others in a social situation. No wonder that he was a success with women.

How Mileva managed to ensnare Einstein is an open question. Perhaps the young man who had resolved in childhood to lead a life of emotional independence, totally devoted to his quest for the meaning behind existence, wished to have what he thought would be a relaxed and cozy relationship with someone he could respect and with whom he felt at ease. Mileva was certainly the type who would do her best to hold a marriage together through the worst of times and she was reliable and strong-willed. Einstein seems to have broken down many of her emotional barriers during the long years of their courtship and, if there was not a particularly great physical attraction between them, they were good friends. During the early years of their relationship at least, there seemed to be an empathy between these two personalities, one outwardly extrovert and easy to be with, but isolated below the surface, and the other outwardly cold but possessing hidden depths of emotion.

While Einstein forged his relationship with Mileva, college was still the first priority in his life. In early 1900, the final examination for his diploma course was approaching fast and he was ill-prepared.

It was Marcel Grossman who was in large part responsible for saving his friend's college career, for he enabled Einstein to catch up with what had been missed through infrequent attendance at lectures. Grossman passed on his beautifully transcribed lecture notes to Einstein and regularly filled him in on what had been covered at college that day. Einstein then had to swot up the material. The effort left its mark. Of this period he said:

> There were altogether only two examinations; aside from these, one could just do as one pleased. This was especially the case if one had a friend, as did I, who attended the lectures regularly and who worked over their content conscientiously. This gave one freedom in the choice of pursuits until a few months before the examination, a freedom which I enjoyed to a great extent and have gladly taken into the bargain the bad conscience connected with it as by far the lesser evil . . . It is, in fact, nothing short of a miracle that the modern methods of instruction have not yet entirely strangled the holy curiosity of inquiry; for the delicate little plant, aside from stimulation, stands mainly in need of freedom; without this it goes to wrack and ruin without fail. It is a very grave mistake to think that the enjoyment of seeing and searching can be promoted by means of coercion and a sense of duty. To the contrary, I believe that it would be possible to

> rob even a healthy beast of prey of its voraciousness, if it were possible, with the aid of a whip, to force the beast to devour continuously, even when not hungry, especially if the food handed out under such coercion were to be selected accordingly.[9]

Despite his deep distaste for enforced learning, Einstein realised that the work had to be done and that if he were to have any chance of future scientific success, no matter how humble his employment, he had to pass the final examinations.

Of course, he did succeed and, with his two friends Grossman and Besso, obtained his diploma from the board of examiners on 28 July 1900. Out of a maximum of 6 marks he obtained 5 for theoretical physics, experimental physics and astronomy, 5.5 for theory of functions and 4.5 for a diploma paper, giving him a final mark of 5 out of 6.

The friends immediately went out to celebrate their success. That evening Einstein allowed himself to revel in his achievement and to be patted on the back for managing to pass with flying colours a series of examinations he loathed having to take. However, there were dark times ahead. Success in examinations was one thing but now he had to face the realities of life and support himself. After his approach to life at the ETH, he was not to find this such a simple task.

As we shall see in Chapter Five, the work Einstein eventually found, though congenial, bore little relation to the ambitions he had sketched out as a schoolboy. But even before he obtained an academic post, and as soon as he had completed his studies at the ETH, Einstein began to make original contributions to science, and from the outset these bore his unique stamp.

Chapter Four

Early Works

Although Einstein had completed his studies at the polytechnic successfully, he was unable to obtain a teaching or research position there or at any other university in the next few years. During 1901, he found some temporary work as a teacher at high-school level, and he continued to pursue his scientific interests under his own steam, with the aid of the library at the polytechnic. He hoped to produce an original contribution to science of a sufficiently high standard to qualify him for the award of a PhD from the University of Zurich (the polytechnic did not award the PhD degree then), and as early as August 1901 he submitted an investigation of the kinetic theory of gases to the university. It was not accepted as a satisfactory thesis, although it formed the basis of Einstein's third published scientific paper. Others followed, as Einstein moved to Bern and took up the famous job as a patent officer in 1902 (the move and job are discussed in Chapter Five); eventually, the scientific work that he carried out in his spare time from that job would, indeed, lead to the award of a PhD. As we shall see, by the time he was able officially to call himself 'Herr Doktor Einstein', the degree was almost an irrelevance, since Einstein had by then begun to publish papers of such significance that the scientific world could not fail to sit up and take notice, whatever the paper qualifications of the author. But it is still intriguing to see how that early work developed – and the thesis that was eventually accepted as worthy of a PhD, in 1905, was a sufficiently impressive piece of work to have stood the test of time in its own right, even though it does not deal with the topics that made Albert Einstein a household name.

First steps

Einstein's first two scientific papers, published in 1901 and 1902, deal mainly with the nature of the forces between molecules, and attempt to explain, for example, the nature of liquid surfaces. The whole basis of these papers is just plain wrong. Einstein based his calculations on the idea that the forces between molecules obey a universal law, similar to the law of gravity. We now know that these forces, although they are essentially electrical, depend, among other things, on the actual size of the molecules, and that the nature of the attraction between molecules in a liquid or gas (or even a solid, come to that) is much more complicated than Newton's law of gravity.

But even so, these papers are interesting. First, they show how from the very beginning of his scientific career Einstein was trying to find unifying themes in physics – in this case, trying to find similarities between gravity and intermolecular forces. This was a theme that was to recur throughout his life, and to haunt the largely fruitless labours of his final years. Secondly, the fact that Einstein was thinking along these lines at all right at the beginning of the twentieth century shows how the attitude of most physicists towards the notion of molecules and atoms had, by that time, diverged from that of some chemists, and points the way towards one of the three great contributions to science that Einstein was to make just a few years later, in 1905.

Physicists had always been one step ahead of the chemists as far as atomic theory was concerned. Isaac Newton wrote, in his *Opticks*, of matter being made of 'primitive Particles . . . incomparably harder than any porous Bodies compounded of them; even so very hard, as never to wear out or break in pieces', and in 1738 the Swiss mathematician Daniel Bernoulli described how the behaviour of a gas could be explained in terms of the motion of many tiny particles, colliding with one another and with the walls of their container. In the nineteenth century, physicists such as James Clerk Maxwell seem to have taken the notion of atoms for granted; but the chemists struggled for decades to come to terms with the idea.

The first steps towards the modern understanding of atoms in chemical terms were made by the English chemist John Dalton, in the first decade of the nineteenth century. This understanding envisages an atom as the smallest unit of an element which can take part in a chemical reaction. Atoms of hydrogen and oxygen, for example, can combine to form molecules of water, and molecules of water can be broken down into atoms of oxygen and hydrogen; but the atoms of hydrogen and oxygen cannot themselves be broken up into anything smaller, so they are more elementary than the molecules. The chemical investigation of the nature of atoms was handicapped because, we now know, many elements actually come in different varieties. There are atoms of carbon, for example, that have identical chemical properties but different masses (different weights). Such atoms which are chemically identical but physically different (if only in terms of mass) are now known as isotopes of the same element.

Grudgingly, chemists came to terms with the notion of atoms in the nineteenth century. The idea was almost forced upon them, as the only way in which they could explain the nature of chemical reactions, but even late in the nineteenth century it was still widely regarded as a working hypothesis, a way of thinking about things that seemed to work for all practical purposes, but might yet be replaced by a better understanding of the nature of chemical reactions. In 1901, when Einstein was writing his first scientific paper, and happily using the idea of molecules, there was still no accepted, direct proof that atoms and molecules really existed.

That is not to say that there was not strong circumstantial evidence. The most fruitful line of attack stemmed from a brilliant guess made by the Italian physicist Amadeo Avogadro in 1811. He said that if you had a box of a certain volume, and filled it with gas at a certain temperature and pressure, then whatever gas you put into the box at that same temperature and pressure, there would be the same number of particles (either atoms or molecules) in the box. The idea behind this is, essentially, that as far as the pressure on the walls of the box is concerned it doesn't matter what the particles hitting it are made of, only their speed and how often they hit it. Their speed depends on the temperature, and how often they hit it depends on how many particles there are in the box. So if the temperature, volume and pressure are the same, the number of particles must be the same.

Even Dalton did not accept this idea immediately, and it was not until the 1850s that it began to have a big impact on the way physicists and chemists thought about atoms and molecules. But the important thing about Avogadro's hypothesis was that it gave scientists a handle on atoms and molecules. If Avogadro was correct, there were obvious ways in which it might be possible to calculate the number of particles in a box of gas.

In calculations of this kind, the experts generally choose to work in terms of a box of gas at zero degrees Celsius and one standard atmosphere of pressure. The standard volume is chosen so that it would contain the same mass in grams as the molecular weight of the gas compared with the weight of one atom of hydrogen, the lightest element. This is called the 'gram molecular weight'. Such a box would contain 32 grams of oxygen gas, 2 grams of hydrogen (because there are two atoms in each molecule), or the appropriate, unique mass of any other gas. But in every case the number of molecules in the box would be the same – a number now known as the Avogadro constant, or sometimes as Avogadro's number, denoted by N and shown by modern techniques to be just over 6×10^{23} – that is, a 6 followed by 23 zeros. In one cubic centimetre of air under the same conditions, there are 4.5×10^{19} molecules, which gives you some idea of just how tiny molecules and atoms really are.

But this is getting slightly ahead of our story. Einstein's own attempt at calculating Avogadro's number actually came in his PhD thesis – and that didn't appear until he had published three more papers, rather more impressive in their own right than his first two efforts.

Scientific respectability

The other three early scientific papers, published between 1902 and 1904, deal with thermodynamics and the nature of the second law. The papers are flawed by Einstein's lack of access to the kind of facilities he would have had as a university professor, and because he had not read all the work of the pioneers such as Boltzmann and Maxwell, and the American Josiah Willard Gibbs, he sometimes went over the same ground as they had covered a generation before, sometimes rediscovering things that were already known, and sometimes following up alleys that were already known to be cul-de-sacs. But this experience may, in the long run, have stood Einstein in good stead. By working out much of the basis of what is known as statistical mechanics – an approach to thermodynamics that depends on calculating the statistics of the behaviour of large numbers of particles, atoms or molecules – on his own, he gained a thorough mastery over his subject, and was later able to apply these ideas in many new ways. What is more, the kind of statistical techniques that Einstein used initially in his investigation of thermodynamics soon proved to be particularly useful in the new science of quantum mechanics.

Like all scientists, Einstein had to gain familiarity with the tools of his trade before he could begin to apply those tools to the cultivation of new and fertile ground. But, in view of the way his name later became linked with our understanding of time, even one of those cul-de-sacs is, with hindsight, worth a mention.

In one of those early papers, published in 1903, Einstein thought that he had found a proof that the second law of thermodynamics must always hold. The second law, remember, is the one that says things wear out. In modern language, Einstein thought that he had proved that the arrow of time is built into the laws of physics, although he never used that expression himself. But he had failed to appreciate one of the key points in Boltzmann's argument, which jumped off from a comment made by Johann Loschmidt, whom we shall meet again shortly, and was published in 1877, in a paper Einstein cannot have read in 1903. This made it clear that the second law is not an absolute law of nature, but only has a *statistical* validity.

Remember the box of gas, one half full, the other empty, with a partition in the middle? When we discussed this in Chapter Two, we talked of pulling the partition out and letting the gas fill the box, then waiting for a very long time to see if the molecules of the gas might ever move together into one end of the box again. We said that, according to Boltzmann, this might happen, after a very, very long time. It would be as if we watched a pile of bricks, for a time much longer than the age of the universe, and suddenly it *did* assemble itself into a house. During such processes, the flow of entropy is reversed and the second law is violated – in the case of the gas in the box, the gas would get hotter as it squeezed itself into one end of the box. Time, by our everyday standards, would be running backwards.

You can see this most clearly by thinking about all the individual atoms in the box, moving rapidly around and bouncing off one another as they spread out from one end. The tried and trusted laws of mechanics, going back to Newton, tell us that no physical law would be violated if the motion of every particle in the box were instantaneously reversed. To Newtonian physics, a particle moving from right to left at a certain speed has as much right to do so as it has to move from left to right at the same speed. If one trajectory is physically possible, then so is the other. The effect of this on the particles of gas in our box is that if the motion of each particle were reversed they would continue to rush about and bounce off each other, but now retracing, exactly, their paths back into the half of the box where they had come from.

One implication of this is that, as we mentioned in Chapter Two and as Einstein failed to appreciate fully in 1903, it is possible for random fluctuations, in a very large, very long-lived universe, to produce bubbles of low entropy, which then unwind back towards the heat death. This is one way to resolve the apparent conflict between the fact that there is an obvious arrow of time in the everyday world (things wear out) but not in the particle world (trajectories are reversible). As Ilya Prigogine, who was born in Moscow in 1917 but moved to Brussels at the age of ten, has remarked, it seems logical to accept that 'irreversibility is either true on all levels or on none; it cannot emerge as if out of nothing, on going from one level to another'.

But there is another way to resolve the conflict, one favoured by Prigogine and supported by the idea of quantum uncertainty which we shall discuss in Chapter Ten, but which neither Einstein nor anyone else had any inkling of in 1903. This is the possibility that it is the irreversibility of the everyday world, with its obvious arrow of time, that is indeed 'real', and that particle trajectories cannot, after all, be perfectly reversible, so that the same arrow of time also applies on the microscopic level. This is essentially what Einstein was trying to prove in 1903. His attempt failed because, of course, he was using Newtonian mechanics; quantum mechanics would not be invented for another two decades. But when it was, Einstein would be right there in the forefront, breaking new ground along with the other quantum pioneers.

After publishing just five scientific papers, however, Einstein was, by the end of 1904, ready to make his mark on the world of science. The following year, 1905, has gone down in Einstein legend as the *anno mirabilus* when the new genius burst forth in the world of science with three scientific papers on widely differing topics, any one of which might have earned him a Nobel Prize, and all written by a patent officer in Bern who could not even, yet, sign himself 'Dr' Albert Einstein. But those papers didn't appear quite so much out of the blue as the legend would have you believe; one rather important thing tends to get overlooked. That is Einstein's thesis itself, written before those three great papers appeared in print, although, because of the delay caused by submitting it to the examiners, not published until 1906. It has been described by Abraham Pais, the author of the most comprehensive scientific biography of Einstein, as 'one of his most fundamental papers'.[1]

Herr Doktor Einstein

Pais provides a rather neat example of the scientific importance of Einstein's thesis, in its published form. One of the standard ways to determine how useful a scientific paper is is to count the number of times it is referred to in other scientific papers – the number of citations. These citations are collated and published in standard reference texts. Out of all the scientific papers published before 1912, three of the ten most frequently cited in later papers (published between 1961 and 1975) are by Einstein. This may be no surprise – but the papers themselves, to anyone brought up to think of Einstein solely as the father of relativity theory, are. The most frequently cited of the three, as recently as the 1970s, was, in fact, his thesis, and a sequel to this paper is the next most frequently cited of Einstein's early works. The third is one of the great 1905 papers, on Brownian motion, which we discuss in Chapter Six. The paper announcing the special theory of relativity, also published in 1905, doesn't come in Einstein's top three or the overall top ten of citations for papers published before 1912.

Of course, one reason for this is that special relativity became such a standard feature of physics that long before the 1960s and 1970s hardly anyone ever bothered to read, or refer to, the original 1905 paper. Nevertheless, this scientific ranking does make it clear that Einstein's thesis was something special. The reason why the thesis paper has been so widely quoted, so recently, is that it deals with the properties of particles suspended in a fluid, a topic which has much more practical application in everyday life than the special theory of relativity, and has found uses in calculations as diverse as the way sand particles get stirred up in cement mixers, the properties of cow's milk, and the way fine particles of dust and droplets of liquid (aerosols) are suspended in clouds.

But Einstein didn't set about his investigation of the way particles are suspended in fluids because he was concerned about problems involving cement, milk or dirty clouds. As he later told his friend and scientific sparring partner Max Born, referring to the work leading up to his thesis, 'my main purpose for doing this was to find facts which would attest to the existence of atoms of definite size'.[2] In that aim, he was following a well-established tradition, going back almost a hundred years, and in his thesis he came to the very brink of providing the final clinching proof that would at last persuade the remaining doubters of the reality of molecules and atoms – that final proof actually came in the Brownian motion paper we have already mentioned, which followed hot on the heels of the completion of Einstein's PhD thesis, and was really an extension of the thesis work.

In view of the uncertainty about the reality of atoms and

molecules that remained in many scientists' minds even at the beginning of the twentieth century, it is remarkable how many attempts to calculate the sizes of molecules were made in the nineteenth century, how early the first of those attempts were made, and how accurate the answers they came up with turned out to be. To be honest, many of the calculations were based on weird and wonderful notions about the nature of molecules, and some (though by no means all) of the surprisingly accurate answers were obtained as much by luck as anything else. Nevertheless, by the end of the 1880s those physicists who did believe in molecules also reckoned that they knew the sizes of the beasts to within a factor of two, from 100 to 200 millionths of a centimetre in radius. This is a very sensible estimate, in line with modern calculations – the separation between the centres of the two hydrogen atoms that make up a hydrogen molecule, for example, is about 75 millionths of a centimetre (0.75 angstroms), and the overall extent of the molecule, the smallest that can exist, is slightly bigger than this.

We'll mention in detail just two of the nineteenth-century attempts at estimating the sizes of molecules – the first, and the one most like Einstein's approach to the problem.

The first attempt was made in 1816, by the same Thomas Young that proved the wave nature of light using the double-slit experiment. He worked out the size of water molecules, using an ingenious idea based on measurements of surface tension in the liquid.

Surface tension is a kind of elasticity in the surface of a liquid, like water. It is responsible for the way the edges of the water in a glass cling to the glass, lifting the surface slightly up the sides, and for the way in which droplets of water form into nearly spherical shapes, like tiny balloons. The surface tension makes a barrier between the water itself and the air above, so that if you are very careful you can 'float' a steel darning needle on the water surface. The reason for surface tension is that molecules of water attract each other, pulling with those intermolecular forces that Einstein investigated in his very first scientific paper. In the middle of a glass of water, each molecule is pulled more or less evenly in all directions. But at the surface, the molecules are only being tugged on by other water molecules below them and to the sides. They are not tugged on by the molecules of the air above. The result is that the water surface acts like an elastic skin, a molecule or so thick, on top of the bulk of the liquid.

Young guessed that the strength of the resulting surface tension must be related to the range of the forces acting between water molecules, and that this range must represent the size of the molecules themselves – that they had to be almost touching each other before the stickiness of the force of attraction became apparent. Using these arguments and measurements of the surface tension of water, he decided that the diameter of what he called 'particles of water' and what we would now call water molecules must be 'between the two thousand and the ten thousand millionth of an inch'.[3] Since an inch is 2.54 cm, this placed the size of water molecules at between about 5 and 25 thousand millionths of a centimetre, only about ten times too big, and an astonishingly accurate 'guesstimate' for the time, scarcely a year after Napoleon had lost the Battle of Waterloo.

Fifty years later, in the mid-1860s, a very neat attempt at estimating the sizes of molecules was made by Johann Joseph Loschmidt, a German chemist who had been born in 1821. This was the same Loschmidt who pointed out to Boltzmann the importance for thermodynamics of the reversibility of particle trajectories allowed by Newtonian mechanics.

The nub of Loschmidt's approach, which carries over into Einstein's work, is that he used two sets of equations to determine simultaneously two properties of molecules – their size, and Avogadro's number. This is a standard technique, the memory of which may be familiar even to non-mathematicians from school days. If you have one unknown quantity, and one equation in which that quantity appears (like Young's equation involving surface tension and the size of water molecules), you can solve the equation to find the unknown quantity. If you have two unknown quantities, you need two equations each involving both quantities before you can solve the equations to find out both the unknown numbers. With three unknowns, you need three equations, and so on – but both Loschmidt and Einstein stuck, for this particular calculation, with a pair of simultaneous equations, solving them for two unknown quantities.

Loschmidt's calculations involved the average distance a molecule travels between collisions in a gas – the so-called 'mean free path' – and the fraction of the volume of the gas actually occupied by the volume of all the molecules added together. He assumed, like Young, that in a liquid all the molecules are touching each other, which gave him a handle on the volume occupied by all the particles (molecules) in the liquid when they are closely packed together. Then, when the same liquid was heated to become a gas, he knew that the volume of gas actually occupied by the molecules must be the same as the volume of the liquid that had been evaporated, and that the rest of the volume of the gas is simply the empty space that the molecules whiz through.[4] Since he actually carried out his calculations for air, he had to use estimates of the densities of liquid nitrogen and liquid oxygen which were not as accurate as modern measurements, but he still came up with answers to his calculations that stand up very well even today. Loschmidt said that the diameter of a typical molecule of air must be measured in millionths of a millimetre, and he gave, in 1866, a value for Avogadro's number of 0.5×10^{23}.

Using modern data, the mean free path of molecules of air turns out to be just 13 millionths of a metre at 0 °C, and an oxygen molecule in air at that temperature will be travelling at just over 461 metres per second. So it undergoes more than 3.5 billion (thousand million) collisions every second.

Einstein's approach used a similar form of mathematical reasoning to that of Loschmidt, solving two simultaneous equations with the same two unknown quantities in them. But he applied his reasoning not to gases but to solutions, in which molecules of one compound (the solute) are spread more or less evenly through a liquid which is made up of molecules of another compound (the solvent). There is nothing particularly exotic about the solutions Einstein based his calculations on – they were simply solutions of sugar in water. His calculations would, in fact, apply very precisely to the behaviour of a cup of sweet tea.

The starting point for this work was the discovery, made back in the 1880s but still surprising the first time you encounter it, that the molecules in a solution behave like the molecules of a gas. One example of this is a phenomenon known as osmotic pressure.

Imagine a container holding a solvent (just water, in Einstein's calculation), divided into two halves by a barrier which has tiny holes in it that allow the molecules of solvent to pass through, but are too small to allow molecules of a chosen solute (the sugar) to pass through. Now, if you put sugar into onc half of the container, so that there is a solution on one side of the barrier (often called a 'semipermeable membrane') but not on the other side, solvent will flow through the barrier and into the half of the container that holds the solution. The water, in this case, flows *from* the weaker solution, *into* the stronger solution, trying to establish a thermodynamic equilibrium by evening out the concentration of the solution in both halves of the container.

This is exactly equivalent to the way in which gas spreads out from one side of a box to fill the entire box when a partition in the middle of the box is removed. You might think, on first encountering the problem of two different strength solutions separated by a semipermeable membrane, that the 'extra' molecules in the stronger solution ought somehow to force the solute through into the weaker solution. But if that happened, the strong solution would get stronger and the weak solution would get weaker. There would be a more pronounced difference between the two halves of the container, so entropy would have decreased. In order for information to be lost and entropy to increase, in line with the second law of thermodynamics, the strong solution must somehow be made weaker, more like the weak solution, even if that involves a net flow of solute from the weak solution into the strong solution. As a result, the level of solution rises in the side of the container that contains the sugar, and the level drops in the side of the container holding just water. The process stops when the extra pressure of the stronger solution, caused by the weight of the extra height of liquid in that side of the container, is strong enough to stop the flow of solvent through the membrane. The flow of solvent through the membrane is known as osmosis, and the pressure needed to stop the flow is called the osmotic pressure.

The osmotic pressure depends on the number of molecules of solute in the solution – the more concentrated the solution is, the stronger the pressure. And, once again, the sizes of the molecules involved comes into the calculation in terms of the fraction of the volume of the solution that is actually occupied by those molecules. The second equation used by Einstein involved the mean free path of the molecules of the solute, which he related to the speed with which liquid (molecules) diffused through the membrane. Along the way, he had to determine other properties, such as the relation between this diffusion and the viscosity (stickiness) of a liquid, which proved so interesting to engineers investigating cement, milk and all the rest.

In the thesis itself, Einstein found a value for Avogadro's number of 2.1×10^{23}, with estimates for molecular sizes in the now familiar range of a few angstroms (a few hundred millionths of a centimetre); in the version published in 1906, he was able to improve the calculation by using some new data from more accurate measurements of the behaviour of sugar solutions, which gave him a value of 4.15×10^{23}. And by 1911, Einstein was able to improve the calculation still further in a new paper which gave Avogadro's number as 6.6×10^{23}. By then, this crucial number had been determined reasonably accurately in a dozen different ways, and all those determinations gave very similar values. Each technique independently confirmed the reality of molecules, and there was no longer any doubt that atoms and molecules were real, physical entities.

Einstein's own last major paper using the classical techniques of statistical physics (that is, without using the new quantum physics) was written in October 1910, and concerned the way in which the blue colour of the sky is produced by light scattering from the molecules of the air itself. As far back as 1869, the British physicist John Tyndall had explained that the blueness of the sky might be caused by the way in which small dust particles or droplets of liquid in the air would bounce blue light (which has short wavelengths) around, scattering it to all parts of the sky, while red and orange light (which has longer wavelengths) could pass through relatively unaffected (explaining why sunrises and sunsets are red). Other scientists realised that the scattering must actually be caused by the molecules of air themselves. But it was Einstein who put the numbers in, proving that the blueness of the sky was connected with the existence of molecules and deriving the value of Avogadro's number in yet another way in the process.

Shortly after completing this piece of blue-sky research, Einstein moved to Prague, becoming a full professor for the first time, and beginning his final assault on establishing the general theory of relativity.

But that is getting ahead of our story. Although Einstein did carry out other work in statistical mechanics, investigating, among other things, the nature of specific heat (the way in which a certain amount of heat input to an object will cause a certain rise in its temperature), all these investigations pale into insignificance compared with the three great achievements of 1905 and Einstein's later, towering contributions to the two greatest developments of twentieth-century physics, relativity theory and quantum mechanics. These are themes that we shall concentrate on in later chapters. But first, the *anno mirabilis* itself. Pride of place among the three great papers published in that year should perhaps go to the one that is most taken for granted today, but which follows directly from Einstein's *doktorarbeit*, which has itself been, as we have seen, probably his most unsung major contribution to science. Perhaps the most remarkable feature of all this work, however, is that it was not carried out by an academic secure in a university post, but by an apparently failed academic, working as a civil servant in the Swiss Patent Office.

Chapter Five

Albert Einstein – Patent Officer

Albert Einstein's animosity towards 'Herr Weber' was to cost him eighteen months in the employment wilderness, and had it not been for his natural arrogance and self-assurance, it would perhaps have given him cause to feel regretful that he had not displayed greater diplomacy and respect towards his social superiors. On the contrary, what Einstein saw as Weber's deliberate interference only served to harden his resolve to go his own way.

Out of the group who passed the examination in his class that summer of 1900 and wanted to go straight into teaching or higher education, Einstein was the only one who did not find an immediate teaching position or postgraduate appointment. It was particularly galling for Einstein to see sometimes wholly unsuitable students landing positions perfectly appropriate to his training. Weber, a physicist himself, took on two mechanical-engineering graduates and passed over the physicist Einstein. Of Einstein's friends, Marcel Grossman and Jakob Ehrat were both taken on by ETH professors.

Einstein's bleak employment situation had produced two immediate problems. The first was financial. With the conclusion of his studies, the generosity of his mother's family was exhausted and his own parents were not able to fund him. The second was his acquisition of Swiss citizenship, which he had been striving for since his arrival in the country in 1896. It required that he could prove he was in full employment.

The latter of these difficulties was overcome when Einstein managed to persuade one of his professors, Alfred Wolfer, with whom had enjoyed a friendly relationship, to allow him to work with him for a short time at the Swiss Federal Observatory. Wolfer had just been made the director of the observatory, and had for some time considered Einstein to be a bright and promising student.

Still, Einstein's application for citizenship proved problematical. By his own account, he had to undergo rigorous examination before the authorities were satisfied that he was suitable. He had made his final application well over a year earlier, in the summer of 1899, but it was not until the summer of 1900 that his father was asked for his declaration of approval. This was duly furnished and in February 1901, five years after his first application, Albert Einstein was made a Swiss citizen and given civil rights of the city of Zurich.

Einstein remained a Swiss citizen until his death in 1955 and maintained a lifelong love for the country. When he was an elderly man, living in the academic community of Princeton University, he wrote: 'I love the Swiss because by and large, they are more humane than the other people among whom I have lived.'[1]

Humanity and what he perceived as civilised attitudes were very important to Einstein. What had swayed him in his choice of nationality, apart from the language, was the centuries-long nonmilitary, noninterventionist stance of the Swiss. Switzerland was a nation which played as minor a role in world affairs as the young scientist wished for himself outside the realm of science; an individualistic nation going its own way in much the same way that he wished his own life to progress.

However, in the spring of 1901, he had precious little independence from the world. He was chasing after any reasonable job he heard about and at the same time trying to find his way with his scientific explorations.

One of his first duties as a Swiss citizen was to present himself for the three month military service compulsory for every young man in the country. He was examined and, much to his annoyance at the time, rejected because of flat feet and varicose veins. It appears that the anti-military side of Einstein was overshadowed by what he may have perceived as a personal affront. He always felt better if he could reject authority rather than it rejecting him, for whatever reason.

In the spring of 1901 Einstein returned to his parents' home in Milan and began sending out one letter after another to eminent scientists around the continent, requesting that they consider him as an assistant. As one would expect, all his letters were left unanswered. But on 14 April 1901 he received a letter inviting him to take up a temporary teaching post at a technical school in the town of Winterthur, about twenty miles northeast of Zurich. The position became vacant because one of the teachers at the school was required to do military service and a replacement had to be found for the final term of the year between 15 May and 15 July.

Einstein enjoyed teaching and this, his first real taste of it, passed quickly. He appears to have done well at the school, but was not offered a permanent position there. Come the summer, he was once more unemployed. Then he received another break. He applied for a teaching position advertised in a Zurich newspaper. The job was in a private school in the small town of Schaffhausen. It so happened that the family of one of Einstein's college friends, Conrad Habicht, lived in the town and knew the owner of the school. Habicht was able to pull a few strings for his friend and Einstein was offered the job as private tutor to an English boy, Louis Cohen, who was attending the school but needed extra help with his work. Einstein accepted the position immediately.

With the guarantee of a full academic year at the Schaffhausen school, a reasonable salary and enough spare time to work on his PhD thesis, Einstein's problems seemed solved. Then disaster struck. In July 1901, Mileva informed Einstein that she was pregnant by him.

Einstein's earliest recorded response to the dramatic news was that he would do everything he could to secure a good job and support the three of them. In a letter to Mileva, dated July 1901, he writes:

> But now, rejoice in the irrevocable decision that I have made! I decided the following about our future: I will look *immediately* for a position, no matter how humble. My scientific goals and my personal vanity will not prevent me from accepting the most subordinate role. The moment I have obtained such a position I'll marry you and take you to me without writing anyone a single word before everything has been settled. And then nobody can cast a stone upon your dear head and whoever dares to do anything against you, he'll better watch out! When your and my parents are faced with the fait accompli they'll just have to reconcile themselves with it the best they can. And as my little wife, you can peacefully rest your little head in my lap and will not have to regret the tiniest bit the love and devotion you have bestowed upon me.[2]

From the little documentary evidence that has survived from the time, he seems to have taken the whole thing pretty calmly. In the correspondence between the couple there are occasional mentions of the pregnancy and veiled references to a child they planned to call Lieserl.

But things were not as simple as Einstein had made out. His mother, Pauline Einstein, intensely disliked Mileva and entirely disapproved of the relationship. It is difficult to ascertain the reasons for Pauline Einstein's feelings, but it is safe to assume that she saw Mileva's family as beneath the Einsteins' social standing. Mileva came from peasant stock and her family were relatively poor. Aspiring to higher things, Mileva had been accepted at the ETH, but the fact that she was taking a scientific course may have suggested to Frau Einstein that Mileva was in some way unconventional and peculiar. It is also possible that Mileva's personality contributed to the animosity. However, from surviving letters, it is clear that Pauline Einstein did not like the idea of the relationship between Mileva and her son long before the two women met.

Things got so bad at one point that the Einsteins actually wrote to the Marics to inform them of their feelings about the relationship between their children. Little did the Einsteins know that, at the time, their son's fiancée was over six months pregnant.

Mileva was quite naturally upset about this letter. She went to Schaffhausen to talk to Albert about it. Unfortunately, Maja was visiting her brother and Mileva had to wait a couple of days before the father of the child she was carrying could get away to see her. Although Maja was on friendly terms with Mileva, she was not in on the secret.

The Marics, on the other hand, saw Einstein as an entirely suitable young man for their daughter. They were very hospitable to him when he visited them at their home and from what Mileva says in her letters, her parents were supportive of the relationship. So, the couple eventually confided in Mileva's parents about the pregnancy. They had little choice; during the autumn and winter of 1901 she made frequent visits home and before too long her condition would have been quite apparent.

Once again the Marics were immediately supportive, though initially Mrs Maric's feelings towards Einstein cooled. This much is evident from a cheeky passing comment of Einstein's in a letter to Mileva dated 17 December 1901, when Mileva was already eight months pregnant: 'Give my kind regards to your old lady and tell her also that I am looking forward to the thrashing she will bestow upon me one of these days.'[3]

Mileva gave birth to a daughter at the end of January 1902. They called the girl Lieserl, just as they had intended from the very beginning. But, despite the original naive wishes of the young couple, they were not able to keep the baby.

What happened to Einstein's first child is not known. The most likely scenario is that Mileva's family looked after the child and that she grew up in southern Hungary. However, after the probable adoption, no further mention was made of her, and all attempts at tracing Lieserl's eventual fate have drawn a blank.

The fact of Einstein's illegitimate child from his first marriage was kept secret for the best part of a century. Yet in this series of events lies the real reason why Mileva has remained a somewhat shadowy figure in the recorded life of one of this century's greatest men.

For some, what has appeared to be a suppression of information about Einstein's first wife has inaccurately been seen as an attempt to cover up her influence on her husband's work.

In recent years, a furious debate has blown up amongst Einstein scholars concerning Mileva Maric's contribution to the creation of relativity. Claims that Einstein's first wife was crucial in the early stages of the theory's development first arose at a meeting of the American Association for the Advancement of Science in New Orleans in 1990. Two Einstein scholars, Dr. Evan Harris Walker and Professor John Stachel clashed over the question and a heated debate followed. Walker supported Maric's case, Stachel strongly opposed it.

Within days the argument had made world headlines, and it was discussed and dissected by journalists and academics alike. It now appears that Walker's claims are unsubstantiated and, at best, far-fetched. Although Mileva studied physics at the ETH, and she and Albert often discussed science in some detail, it is generally agreed by most scholars that Mileva Maric was not in the position to make a significant contribution to the theory of relativity. There is little question that she was a tenacious and hard-working student who helped Albert by checking his calculations. However, it is doubtful that she had the imagination or the deeper understanding of fundamental physics that might have enabled her to come up with such an other-worldly theory as relativity.

Despite the fact that in their correspondence, Einstein would on occasion refer to 'our work' when discussing his latest ideas with his fiancée, Mileva is unlikely to have been more than a moral support for her future husband in his scientific work. Phrases like 'our work' merely show that Einstein wanted to create a closer relationship between them and make Mileva feel that she was sharing in his theories.

Mileva's pregnancy also gives us a plausible reason for why she failed her second attempt to qualify at the ETH. By the time of the examinations she was three months pregnant, her family had not been informed and she had only just imparted the news to the father. She may well have been feeling physically ill during the time she spent revising and emotionally disturbed by the impending chaos she must have seen as inevitable.

*

While the drama of his illegitimate child was unravelling, Einstein was adapting to his new job in Schaffhausen. At first things went well, but within a few weeks cracks began to appear. Little is known of why Einstein left Schaffhausen after only one term, but it may be supposed that what he saw as a correct approach to teaching was not what the school authorities had in mind. Einstein always preferred a liberal, open approach to learning and from the correspondence of the time he had applied his beliefs against the style of the school principal.

Despite the upheaval in his personal life, it would seem that the end of 1901 was a time of growing optimism for Einstein. There were two reasons for this. The first was his submission of a PhD thesis to the University of Zurich. The second was the prospect of a permanent job, thanks to an opening created by his old college friend Marcel Grossman.

Grossman was aware of Einstein's employment difficulties and was scornful of the fact that the establishment saw fit to reject a man with such obvious talents. Grossman's father was a friend of the director of the Swiss Patent Office, Friedrich Haller, and through this connection the Grossmans were able to lay the way open for Einstein to make contact with Haller.

By the end of 1901, things were looking up. Haller wrote to Einstein suggesting that he should apply at once for a newly created position at the Patent Office. A few weeks later the position of patent officer (second class) was advertised. Einstein applied and in due course was invited for an interview in Bern.

The interview with Friedrich Haller went well, despite Einstein's obvious lack of technical experience. At the interview, Einstein was asked to give his opinion of a number of patents put before him. Although as a youth he had been interested in building models and playing with mechanical devices, it soon became clear that his theoretical ability far outshone his technical knowledge and engineering skills.

Perhaps Haller saw something he liked in Einstein's character and manner, or maybe it was because of his friendship with the Grossmans; whatever prompted it, he overlooked Einstein's shortcomings at the interview. Without making any guarantees for employment in the near future, he made it clear that Einstein was in with a very good chance.

Einstein left the interview brimming over with confidence. In a burst of youthful optimism, he immediately handed in his resignation at the school in Schaffhausen and by the beginning of February 1902 had moved to Bern.

Upon his arrival, Einstein entered an entirely new phase of his life. The hoped-for job was not actually offered until June, which meant that in purely financial terms he would have been far better off remaining in Schaffhausen, where he had been earning a reasonable salary. However, despite the material privations his decision to leave had brought him, those first few months in Bern were a time of great intellectual freedom and creativity.

His first home in Bern was a single-roomed flat at Gerechtigkeitsgasse 32. Einstein wrote to Mileva as soon as he arrived there and sent her his first impressions of the city and his new abode:

> It's delightful here in Bern. An ancient, exquisitely cozy city, in which one can live exactly as in Zurich. Very old arcades stretch along both sides of the street, so that one can go from one end of the city to the other in the worst rain without getting noticeably wet. The houses are uncommonly clean, I saw this everywhere yesterday when I was looking for a room . . . I have a large, beautiful room with a very comfortable sofa. It only costs 23 francs. This is not much, after all. In addition, 6 upholstered chairs and 3 wardrobes. One could hold a meeting in it . . .[4]

To make ends meet, Einstein placed an advertisement in a Bern newspaper offering his services as a private tutor. Within days he had taken on his first pupil, a young Romanian student from Bern University called Maurice Solovine.

Many years later, in an account of his time in Bern, Solovine recalled his first encounter with Einstein when he arrived at the latter's room:

> Having entered and taken a chair, I told him that I studied philosophy but that I wanted to study physics a little more thoroughly to gain a real knowledge of nature. He confided in me that he also, when he was younger, had a strong taste for philosophy, but the vagueness and

arbitrariness which reigned there had turned him against it, and that he was now concerned solely with physics. We talked for about two hours on all sorts of questions and we found we had similar ideas and were drawn towards one another. As I left he accompanied me downstairs and we talked for another half hour in the street before an appointment was made for the following day.[5]

Before long a second student had joined them, Conrad Habicht, an old friend of Einstein's from his days in Zurich. Initially Einstein, who was only a couple of years older than the other two, charged them two francs each per lesson and he taught them together. Within a few weeks, the lessons had completely lost their structure and turned into informal discussions. A close friendship grew up between the three of them and they would take themselves off into the hills and go for long rambles in the neighbouring countryside, talking physics and philosophy all the way there and back again.

The three men gave themselves the tongue-in-cheek title of the Olympia Academy. They were often joined by others, who may be considered associate members of the circle. There was Lucien Chavan, an electrical engineer who lived near Einstein and Michelangelo Besso, Einstein's closest friend at the ETH, who had also moved to Bern. It seems that it was with Besso that Einstein forged the deepest relationship, discussing with him many of his early ideas about relativity. The two men often talked late into the night, Einstein bouncing ideas off his friend and in return receiving original new ways to view his embryonic theories. Besso was also responsible for introducing Einstein to the term 'Brownian motion' to describe the random motion of tiny particles in a fluid, caused by the bombardment of fast-moving molecules, a subject of one of Einstein's great papers of 1905.

Einstein acknowledged the debt to his friend in two ways. In 1904 he secured Besso a position at the Bern Patent Office and a year later he dedicated his first relativity paper to him. Further links were established when Besso married Anna Winteler, the sister of Paul Winteler, who became Einstein's brother-in-law when he married Maja Einstein.

In the summer of 1902, Einstein's days of unfettered freedom came to an end, for on 16 June he was officially appointed to the Bern Patent Office and a week later he began work. Einstein's job at the Patent Office suited him perfectly. It was not at all demanding, despite his initial lack of technical knowhow. Haller appointed him technical expert (third class) rather than second class because he thought that Einstein needed to develop his technical knowledge. However, he very quickly adapted and made good any shortcomings.

Einstein had the right sort of mind to tackle the job he was now presented with. He had always enjoyed solving practical puzzles and as a youth he had displayed an interest in experiment before it was driven out of him by the regimentation of his education. The job involved reviewing proposals for inventions sent in to the Patent Office by inventors seeking a patent for their devices. It was Einstein's responsibility to decide whether the idea of the inventor would actually work and to describe in detail any faults he might find with the proposal.

Many commentators have remarked that they found Einstein's ability to reach an almost instantaneous and invariably correct decision about the usefulness of an idea unnerving. He was said to have had an amazing facility to see a problem simultaneously from a number of angles and to judge in advance all the possible consequences of a particular decision. There is little doubt that this was a natural ability and that his experience of working at the Bern Patent Office served to hone this talent into a devastatingly efficient intellectual tool which was to be of inestimable value to him in his scientific endeavours.

There were other advantages to be gained from his employment. His salary of, initially, 3,500 francs per annum, although not a princely sum, was to solve his pressing financial difficulties. It enabled him to break his material ties with his family and to find a much greater independence.

Perhaps the greatest benefit from the position was the fact that he had a good deal of free time, during which he could concentrate on his scientific work. The job offered lifelong security. In effect he could lead a dual life of dedication to his job while he was there and total commitment to his scientific pursuits in his spare time.

During the course of the next six months, Einstein's life went through a number of upheavals. In October, Hermann Einstein suffered a heart attack.

Albert travelled from Bern to Milan to visit his dying father. Hermann Einstein ended his struggles by asking everyone to leave his room so that he could die alone. According to close friends, it was a moment which Albert could not recall without feelings of great guilt.

On his deathbed, Hermann Einstein finally consented to his son's marriage to Mileva. This decision was not sanctioned by Pauline Einstein, but now that Albert had a respectable job, with promotional prospects, a reasonable salary and his own scientific work well in hand, there was nothing stopping him from doing what he wanted to do.

Albert and Mileva were married on 6 January 1903 in Bern. The reception was a quiet, low-key affair with just a few friends in attendance. The couple could not afford a honeymoon, but after the service, the party moved on to a restaurant for the wedding breakfast before the newlyweds returned to their new flat at Kramgasse 49, near the city centre. Arriving there late in the evening, Einstein discovered that he had lost his key and a disgruntled landlord had to be woken up to let them in.

According to some accounts, Einstein began married life with some trepidation. Others have questioned why he married Mileva in the first place. Aside from the rather obvious fact that the couple were in love, there were a number of understandable reasons for getting married when they did.

It has always been known that Einstein craved relief from domestic responsibility. Although he played his part at home, he really needed to be looked after. In many ways this is hardly surprising. For many years he had to do two jobs simultaneously. He was earning a living at the Patent Office, and his spare time was occupied by his scientific work, including reworking his PhD thesis, which had been rejected on its first submission in 1901.

Although Einstein considered working on science a great pleasure and the effort came easily to him, it was still work and therefore time-consuming. He needed someone to keep a home for him. What is surprising is that he could visualise the Mileva Maric he had met at the ETH as a housewife. Surely she had appeared far too independent and self-motivated for such a role?

Between their time at college and their marriage in 1903, Mileva had undergone as much of a change in her outlook as her future husband had done. By 1903, she had lost much of her early interest in science. Einstein later told friends that, by this stage in their lives, she showed almost no interest in what he was doing in physics; this attitude eventually contributed to the break-up of their marriage.

The reason for this change in Mileva may never be fully determined. Perhaps it was her failure to pass her degree at the ETH and the pressures of an illegitimate pregnancy that altered her outlook. She may also have been influenced by the obvious antagonism shown by the Einsteins, which encouraged her to adopt a more conventional stance for a young woman of the time. Whatever the reason, by 1903, Mileva Einstein was slipping into a comfortable role as the wife of a young civil servant. As she rapidly mellowed, her ambition gave way to a softer personality, and it is evident that she too was in love.

Another practical factor in their decision to marry was that they had both had enough of Pauline Einstein's constant attempts at interference in their relationship. Although by marrying they would not put an end to her dislike of Mileva, at least they would be totally independent of parental control. The long periods of separation that Albert and Mileva had to endure during their courtship had also placed a strain on their relationship. Albert and Mileva's early marriage was therefore not so surprising.

The beginning of 1903 heralded the most broadly creative period of Einstein's life. While he maintained what appeared to the outside world to be a humdrum existence as a minor civil servant, the work he did in his spare time would revolutionise physics for the rest of the century. For the next three years Einstein sat in his tiny apartment in Bern or in the park, a short walk from his desk in the Patent Office, and developed a series of scientific papers which would lay the foundations for his truly monumental work of the following decade.

What makes Einstein's achievement even more astonishing is the fact that it was accomplished by a man who had not yet been granted a doctorate, a man who was totally isolated from the rest of the scientific community and who had been rejected even for the most lowly of academic positions.

He was not a graduate of one of the great universities; instead he had passed out of a technical college with a teaching diploma. His resources were extremely limited – the library at the Patent Office, scientific periodicals and his own books. He had very little contact with other scientists, and although the Olympia Academy, which had survived the marriage and regular employment of a founder member, was a rewarding and stimulating intellectual backdrop, it could not substitute for the academic intensity of a university environment.

Einstein's emergence into the scientific spotlight came about via a combination of perseverance and word of mouth. Aside from the publication of a couple of minor papers on the subject of intermolecular forces, his first really serious attempts at publication in an important journal came in 1905, when he published four papers on apparently diverse branches of physics in *Annalen der Physik*. Three of the four turned out to be papers of such enormous originality that the scientific community had to sit up and take notice.

Very soon after their publication it became clear to other physicists that Einstein's papers were complete entities, whole in concept, waterproof in design and revolutionary in their scope.

During the three years it had taken him to develop the theories put forward in the 1905 papers, Einstein had been simultaneously working on his PhD dissertation. This he finally completed in April 1905.

With his revolutionary publications causing a stir in the worldwide scientific community, the University of Zurich accepted his dissertation during the summer of 1905 and awarded Einstein his PhD.

At this time, Einstein's home life had fallen into a routine pattern of contentedness and cozy domesticity. On 14 May 1904,

Mileva gave birth to their second child, a boy they named Hans Albert. Little more than two years after the birth of Lieserl, the couple had a child they could call their own and for whom they could provide a home. At the Patent Office, Einstein's work was viewed with enthusiasm by his superiors and Haller was pleased to note that his decision in employing the technically underqualified young man put forward by his friend Herr Grossman had paid off. In September 1904, Einstein's annual salary at the Patent Office was increased from 3,500 francs to 3,900 francs.

The family went on holiday to Belgrade in the summer of 1905 and there were trips to the Oberland and visits to Albert's uncle Cäsar and his friends the Wintelers in Aarau as well as college pals in Zurich. The rift with Pauline Einstein never healed and Mileva was never accepted as part of the family, but the young Einsteins with their newborn boy visited the Maric family as frequently as the household economy would allow.

Einstein had other distractions and pursuits during those marvellously productive years 1903–05. He still played the violin, finding it an even more useful form of stress relief and intellectual relaxant than in his student days in Zurich. He also took up sailing, a new hobby which would become as important to him as playing the violin. For the rest of his life Einstein enjoyed no better way of relaxing and letting his thoughts wander than to take a small dinghy out on the river or lake and spend an afternoon alone on the water. In later years, when he was an internationally famous scientist and was constantly being visited by eminent men and women from the worlds of science and politics, he would often exasperate his second wife by not being at home to receive a guest because he had taken his boat out on the water for the afternoon.

After the great achievements of 1905, it was obvious that Einstein should be found an academic position. Although he worked alone and within the realms of theoretical physics, he realised that he would benefit from being within an academic environment. But nearly two years passed after the publication of his ground-breaking papers before the first academic opportunity arose. This time it was Professor Alfred Kleiner, a long-established supporter of Einstein and director of the Zurich Physics Institute, who created an opening for the young scientist. Kleiner wanted Einstein on his staff in Zurich, but because of a peculiarity of the Swiss education system, his appointment was not possible straight away.

According to the rules of the educational establishments of a number of European countries at the beginning of the century, an individual could not be appointed to a professorship without having served an apprenticeship as a *Privatdozent*. A *Privatdozent* was required to teach a minimum of a few hours per week to a small group of students. It was not a faculty position and it did not come with a salary. Students paid for the lessons and the *Privatdozent* was paid out of the fees. Thus the *Privatdozenten* earned according to the number of students in their classes. It was an odd system but provided the only avenue into a real academic position at many of the universities of Europe.

Einstein was finally appointed *Privatdozent* at the University of Bern in February 1908. Even this tiny step up the academic ladder had been fraught with problems. In his first application he had overlooked the inclusion of a *Habilitationsschrift*, an unpublished scientific article written by the candidate. After weeks of deliberation this was finally made good and the appointment made.

Einstein was still happily working full-time at the Patent Office. In fact, almost two years earlier, he had been promoted to technical expert (second class), with an accompanying salary of 4,500 francs. As a consequence he had to conduct his lectures at rather unsociable hours, usually on Saturdays or on Tuesday mornings between 7 and 8 a.m. Not surprisingly, very few students turned up to Einstein's first lectures. He began the winter term of 1908–09 by delivering a course of lectures on 'The Theory of Radiation' to only four students. This number fell to a single man by the end of the following term, and often Einstein would end up lecturing to Besso and his own sister Maja who was by this time at the University of Bern, studying for her PhD in Romance languages, a subject about as far removed from the theory of radiation as one could get.

Meanwhile, Einstein's reputation took another great leap when one of his former mathematics lecturers from the ETH, Hermann Minkowski, came up with a mathematical interpretation for the theory of relativity, which gave it an even greater degree of acceptance within the scientific community.

Minkowski was a highly respected mathematician of the first order. He had been successful from an early age. Winning the prestigious Paris Prize for mathematics in 1882, at the age of eighteen, he had gone on to work in Gottingen, the great centre for mathematical development at the turn of the century. In a single popular lecture, 'Space and Time', delivered to the Gesellschaft Deutscher Naturforscher und Ärzte in Cologne on 2 September 1908, and in a paper called 'Basic Equations for the Electromagnetic Phenomena in Moving Bodies' published in the *Gottinger Nachrichten* in 1907, he presented an interpretation of the special theory of relativity in terms of geometry. This pictorial way of understanding the implications of the theory accelerated its spread among scientists.

The collaboration between Einstein and Minkowski proved short-lived. While the younger man was beginning his second term as a *Privatdozent*, in January 1909, Hermann Minkowski died in hospital after developing peritonitis. We can only be left to wonder at what great things might have been achieved by Einstein the physicist and Minkowski the mathematician, working together. Legend has it that Minkowski said on his deathbed: 'What a pity that I have to die in the age of relativity's development.'

Einstein, the patent officer and part-time, unsalaried lecturer, was now being fêted by academic centres around the world. In July 1909 he was awarded his first honorary doctorate by the University of Geneva. Among others awarded this honour at the same time, at a ceremony to celebrate the 350th anniversary of the university, were the discoverer of radium, Marie Curie; the German chemist Wilhelm Ostwald (to whom, only eight years earlier, the young Einstein had written letters begging for a job); and the great Belgian chemist Ernest Solvay. Einstein almost failed to make the celebration because he had thrown away the letter sent by the university authorities, thinking it was a circular. It was only after a second letter was sent, enquiring whether he would accept the honour, that the mistake was made good.

Towards the end of the same summer, in September 1909, Einstein was invited for the first time to deliver a paper. The occasion was a conference held in Salzburg by a German society, Naturforscher Gesellschaft, the Association of Scientists.

This, amazingly, was the first physics conference Einstein had ever attended. It was a grand affair, with many of the leading scientists of the time in attendance. The delivery of his paper would be a far cry from his lectures to a small group of friends in Bern and would, for the first time, put the thirty-year-old scientist under the scrutiny of his colleagues.

By all accounts Einstein's talk went extremely well. Typically, he did not decide to speak on a safe subject or limit his discussion to well-established areas of his work. Instead, he delivered a revolutionary lecture entitled 'Our Views on the Nature and Constitution of Radiation', which presented to the world the bizarre concept of the wave-particle duality of light and the equation which within a few short years would become an icon of science, interpreted and misinterpreted the world over: $E = mc^2$

While all this excitement over his scientific activities continued, Einstein's unusual position as a minor civil servant and *Privatdozent* had at last begun to change.

In 1908 a new chair of theoretical physics had been established at the University of Zurich. Once again Professor Kleiner put Einstein forward as the most suitable candidate for the position. Einstein was lucky to receive Kleiner's support because some time earlier the two men had had a partial falling-out. The story goes that Kleiner had visited Einstein during one of his lectures as a *Privatdozent* and had commented on the fact that it seemed to him that Einstein was lecturing at an inappropriately high level. Einstein had taken exception to this criticism and had snapped back: 'I don't demand to be appointed a professor at Zurich.' Undoubtedly, Einstein was feeling frustrated by the red tape surrounding his deferred appointment to a serious academic position, but his impatience almost cost him dear.

Despite the clash between the two men, Kleiner went on to

recommend Einstein for the newly created position in Zurich. There were further unexpected hurdles to overcome before Einstein was finally to be granted a professorial appointment.

The decision whom to appoint was made by the Zurich Board of Education. The selection of Einstein should have been a foregone conclusion. He was already established as a great theoretician and revolutionary scientist, respected by the entire scientific community. As well as this, he was recommended to the board by no less a figure than the director of the Zurich Physics Institute. However, owing to political interference, things did not go according to plan.

The majority of the board were Social Democrats. A rival for the post was one of Einstein's friends from college days, Friedrich Adler, who was now Kleiner's assistant and a *Privatdozent* at the university. He also just happened to be the son of the founder of the Austrian Social Democratic Party.

Adler senior had sent his son to Zurich to study physics in an attempt to prevent him becoming involved in politics. Ironically, in 1916, Friedrich Adler would become as politically embroiled as it is perhaps possible to be when he was arrested and imprisoned for assassinating the Austrian prime minister, Count Stürgkh.

In the event, the Zurich Board of Education passed over Einstein in favour of the far less able Friedrich Adler. However, that was not to be the end of the matter. Adler was a fanatically honest individual. When he discovered that the main reason for his appointment had been the political influence of his father, he made it very plain that he did not want the position offered by the university.

After lengthy discussions and persuasions, Adler got his way and the board reversed their decision. Einstein was invited to return to Zurich to meet with Kleiner and he was offered the new post.

Pleased with the outcome, Adler told the Board of Education:

> If it is possible to obtain a man like Einstein for our university, it would be absurd to appoint me. I must quite frankly say that my ability as a research

> physicist does not bear even the slightest comparison to Einstein's. Such an opportunity to obtain a man who can benefit us so much by raising the general level of the university should not be lost because of political sympathies.[6]

It appears that Einstein himself was not particularly disturbed by the initial decision of the board and had already begun to look around for schoolteaching positions. When the reversal of the decision came, he took it with the same equanimity as the earlier rejection. After settling the problem of his salary, which was initially set at a ridiculously low level by the university, the appointment was confirmed. His salary was set at exactly that paid by the Bern Patent Office, 4500 francs per annum.

On 6 July 1909, Einstein handed in his resignation as technical expert (second class) and became a fully employed academic for the first time in his life. He was never to look back. From that date onwards, Einstein's career went from strength to strength and never again would he be passed over by an institution. In fact, before another decade had passed, he would become the most sought-after scientist in the world, a man any appointments board would bend over backwards to acquire.

But before we consider how this came about, we should examine in detail the work which brought him to this point, the phenomenal creative outpouring which constituted his *annus mirabilis*.

Chapter Six

The Annus Mirabilis

Einstein completed his thesis at the end of April 1905, but he did not formally submit it to the University of Zurich until 20 July. The 21-page paper was soon accepted, although according to Einstein himself he was initially told that the thesis was too short; he added one sentence before resubmitting it, when it was promptly accepted.[1] It isn't clear why Einstein waited from 30 April until 20 July before submitting the thesis in the first place. It may have been that the later date fitted with the administrative routine of the university; or it may well have been that Einstein was simply too busy with his outburst of creativity in 1905 to get around to sending the thesis off any sooner.

It was not as if Einstein sat back with a sigh of relief and took a rest from his thinking about the nature of atoms and molecules once the thesis was completed (and, remember, he was still holding down the day job at the Patent Office). Just eleven days later, on 11 May 1905, Einstein's classic paper on Brownian motion arrived at the offices of the journal *Annalen der Physik*. This was one of three papers by Einstein published that year in the same volume of the journal, which make Volume XVII of the *Annalen* a collector's item today; a slightly extended version of the thesis paper can be found in Volume XVIII, which is not so highly prized by collectors.[2]

The dance of the molecules

Brownian motion takes its name from the Scottish botanist Robert Brown, who lived from 1773 to 1858, and discovered the phenomenon while studying pollen particles in 1827. Brown was using a microscope to look at grains of pollen, which typically had diameters of no more than half a hundredth of a millimetre (0.5×10^{-2} mm), immersed in water. He discovered that the tiny grains move around in a jerky, agitated fashion, zigzagging erratically to and fro. This motion has nothing to do with currents flowing in the water, and seemed to Brown to 'belong' to the particles themselves. He thought at first that this motion might be a sign that the pollen grains were alive, swimming about in the water in this strange fashion, but soon found that he could also see the same behaviour not only with old pollen that had been dried out and preserved for more than twenty years, but also with microscopic particles of just about anything – Brown studied this motion in a wide variety of clearly nonliving particles, including manganese, nickel, arsenic, coal tar and gum resin.

It quickly became clear that microscopic particles of anything suspended not only in water but in any liquid carry out the agitated dance that soon became dubbed Brownian motion. Indeed, particles suspended in air (such as smoke particles) will do the same trick.

In the 1860s, several physicists independently suggested that Brownian motion might be caused by the constant, but erratic, bombardment by the molecules of the fluid in which the small particles are suspended. The idea was not immediately accepted, partly because some scientists assumed that each little movement of a particle – each 'zig' or 'zag' – must correspond to its being hit by a single molecule, and that would have meant that the molecules themselves could not be as small as calculations like those of Loschmidt were already beginning to imply. But at least one nineteenth-century scientist, the Frenchman Louis Georges Gouy, suggested that the motion might be better explained as a statistical effect, with each kick given to the suspended particle being the result of a large number of more or less simultaneous impacts from many molecules. If the particle is constantly being bombarded from all sides, but the bombardment is slightly stronger at any one instant from one direction, then the particle will move accordingly.

But Einstein knew very little of any of this early in 1905

(just as he knew very little of the detailed history of statistical mechanics in 1905) when he turned his attention to the dance of the molecules. In fact, Einstein started out from the other end of the problem. Having established to his own satisfaction the reality of the existence of molecules (in large measure through his thesis work), he went on to consider how the movement of molecules might make itself visible at a level accessible to the technology of the time. In a very real sense, Einstein *predicted* Brownian motion, rather than explaining it, even though he did his work almost eighty years after Brown made his first observations of the phenomenon.

The opening sentences from the 1905 Brownian motion paper provide the following explanation of Einstein's approach:

> In this paper, it will be shown that according to the molecular-kinetic theory of heat, bodies of microscopically visible size suspended in a liquid will perform movements of such magnitude that they can be easily observed in a microscope, on account of the molecular motions of heat. It is possible that the movements to be discussed here are identical with the so-called 'Brownian molecular motion'; however, the information available to me regarding the latter is so lacking in precision that I can form no judgement in the matter.[3]

Einstein's caution in linking his theoretical prediction of visible motion of particles suspended in a liquid with the already observed Brownian motion extended to the title of the paper, which was 'On the Motion Required by the Molecular Kinetic Theory of Heat of Particles Suspended in Fluids at Rest'. Nothing could make it more clear that Einstein did not set out to explain Brownian motion, but had worked out, on theoretical grounds alone, the way particles ought to behave. This is a powerful insight into the way he worked – as a theoretical physicist with a great intuition for the way the world works, who started out from fundamental laws and principles and worked from these out to the implications for the world around us.

But the reason why it was Einstein's paper that convinced

scientists, firstly, that Brownian motion is caused by the dance of the molecules and, secondly, that therefore molecules really must exist, was that his calculations involved precise numbers and equations. It is one thing to say, as some nineteenth-century scientists did, that Brownian motion might be explained, in general terms, as the result of the impact of molecules on the suspended particles. It is quite another to *calculate*, as Einstein did, the precise statistical nature of the impact of very large numbers of molecules with suspended particles, and to use that calculation to predict the precise nature of the zigzagging Brownian motion that would result. A good scientific theory always has to make quantifiable predictions that can be tested by measurements and experiments.

These particular predictions could be tested, once Einstein had made them, by studying and measuring the way the particles moved, and when those measurements turned out to match precisely the predictions of the equations Einstein had developed, the reality of molecules and atoms could no longer be doubted.

It was another example of Einstein's familiarity and skill with statistical mechanics. He made the basic assumption that the suspended particles (we'll call them 'Brownian' particles, even though Einstein wasn't sure of this in May 1905) are being constantly bombarded from all sides by the tiny molecules of the fluid, and that it is equally likely for the Brownian particle to get a kick in any direction as a result of statistical fluctuations in the number of molecules hitting it at any instant. You might guess that in that case the particle ought to stay in more or less the same place, perhaps jiggling to and fro a little as it is kicked first one way and then the other. But once the particle has moved a little bit from its starting point, it is just as likely to get kicked further away as it is to get kicked back to where it started. Einstein calculated that the average distance travelled by a Brownian particle from its starting point actually increases as the square root of the time since it was first kicked away from the starting point. This means that, ignoring all the zigs and zags, and measuring only the straight-line distance from the starting point, in four seconds the particle travels twice as far as it does in one second, not four times as far, while it takes 16 seconds to travel four times as far as it does in one second.

This completely novel prediction made by Einstein stems directly from the random statistical nature of the collisions, and it carries over into many other areas of science (it is sometimes known as a 'random walk').

What mattered in 1905, however, was that Einstein produced an equation which linked the number of molecules (in essence, Avogadro's number) and their speed (which depends, according to the kinetic theory, on temperature) with the *measurable* way in which a Brownian particle would drift away from its starting point. He hadn't carried out the measurements himself, and so he didn't have the numbers to put into the equation to calculate Avogadro's number; but he did use the equation the other way around in his paper, calculating that for a particle one thousandth of a millimetre across, suspended in water at a temperature of 17°C, the distance travelled from its starting point in one minute after all the Brownian zigzagging would be just six thousandths of a millimetre, if Avogadro's number is 6×10^{23}.

This choice of Avogadro's number in the calculation was a bit of a lucky guess, but very much in the range of existing estimates in 1905. Soon, though, the microscopic measurements that confirmed the square-root relation between time and distance travelled were being carried out on Brownian particles (it is done by watching the drift of such a particle past fine cross hairs in the field of view of the microscope), and by using those measurements physicists were able to calculate Avogadro's number itself using the new technique.

This work, and the rather short doctoral thesis that it stemmed from, would have been enough to ensure Einstein's acceptance by the scientific community, and get him an academic job if he wanted one. But this was, in fact, the *second* of the three great papers to appear in Volume 17 of the *Annalen der Physik*. The first had been written in March 1905.[4] It was Einstein's first contribution to the new quantum theory, and it contained the work for which he was later to receive the Nobel Prize.

Light and electricity

Quantum physics began with the work of the German physicist Max Planck in 1900, neatly on the threshold of the new century, and it stemmed from studies of light (or, strictly speaking, electromagnetic radiation in general). By the 1890s, physicists knew that the way in which a hot body radiates electromagnetic energy depends on its temperature in a simple, but (at the time) hard to explain, way. Strictly speaking, this simple relation applies only to a perfect radiator, a so-called 'black body' (a black body is, in fact, a perfect *absorber* of electromagnetic energy, but since the laws of physics are reversible that means that it is also a perfect radiator). Most everyday objects are not perfect radiators or absorbers, but the way they deviate from the behaviour of a black body can be understood in principle, and we shall ignore such subtleties here by concentrating on the ideal case itself.

The best example of a black-body absorber is a hollow sphere with a small hole in its side. Any radiation, such as light, which happens to go in through the hole will bounce around inside until it is absorbed by the walls, warming them up a little. It is very unlikely that the light will just happen to bounce out through the hole, so the hole is, in effect, a perfect absorber of radiation, or a black body.

But if the sphere is heated up, the walls inside will radiate electromagnetic energy. This is in fact, we now know, due to the way electrons (the electrically charged particles in the outer parts of an atom) move; a moving electric charge, remember, produces a magnetic field, and under the right circumstances a moving electron can give up energy in the form of an electromagnetic wave. The energy radiated from the walls in this way will bounce around inside the cavity and fill it up evenly, until a beam of energy streams out from the hole. This beam of energy used to be known as cavity radiation, for obvious reasons, but is now generally known as black-body radiation – even though the body doing the radiating may be red or white hot.

Those common expressions for degrees of heat – 'red hot' and 'white hot' – themselves give a clue to the nature of black-body radiation. An object that glows red is cooler than one that glows orange, which is cooler than one that glows bluish white. And we know that red light has a longer wavelength than orange light, which in turn has a longer wavelength than blue light. An ordinary domestic radiator, which radiates heat you can feel but not see, is actually emitting infrared radiation, with wavelengths even longer than red light (so does your body); an object heated beyond white heat will radiate ultraviolet energy, with wavelengths shorter than those of visible light. In other words, the colour of the radiation emitted by a black body is a direct indication of its temperature.

In fact, any object will emit electromagnetic radiation with a spread of wavelengths. The centre of this spread (the peak of the distribution) shifts towards shorter wavelengths as the object gets hotter, but there is always a cutoff in the energy being radiated on *both* sides of the peak, on the long wavelength side and on the short wavelength side. The result is a spectrum described by a curve rather like a child's drawing of a hill with a rounded top, sloping away on either side. The spectrum, known as the black-body curve, follows a precise mathematical law. It is always the same shape, for any black body, whatever its temperature; and in the late 1890s Planck was trying to find the physical reason for that mathematical law.

At the time, this was a major challenge for physicists, because the laws of physics as they were known at the time seemed to imply that *any* object should radiate an *infinite* amount of energy at very short wavelengths. This disastrous theoretical prediction (which is obviously not a good description of the real world) was known as 'the ultraviolet catastrophe'. It comes from the seemingly innocent assumption that electromagnetic waves inside a cavity can be treated in the same way as waves on a guitar string.

Because there are very many waves with very many different wavelengths to be considered, the laws of statistical mechanics have to be taken over from the world of particle physics and applied to the problem. The application of these laws leads to the conclusion that the energy being radiated at any frequency is proportional to the frequency. Frequency is one over the wavelength, with very short wavelengths corresponding to very high frequencies. So all black bodies ought to produce huge amounts of energy at high frequencies, which means at short wavelengths.

Planck spent years working on the problem, following up several blind alleys and making some leaps in the dark that could not really be justified on the basis of nineteenth-century physics, but they got him, in the end, to a new description of what is actually going on when a hot object radiates electromagnetic energy (or, indeed, when a cold object absorbs electromagnetic energy). This is not the place to go into all the details,[5] but in effect Planck solved the problem by cutting the electromagnetic energy up (mathematically) into small chunks. He did not suggest that these pieces of radiation had any physical significance, but thought that what was happening inside a hot object to make it radiate energy allowed energy to be radiated only in pieces of a certain size. This is rather like the way in which the cash dispenser at your local bank will 'emit' money. Provided your account is sufficiently in credit, you can take out any sum you like, as long as it is a multiple of £10. Other amounts of money, such as £47.38, can exist in the real world of your pocket, but the machine will only cough up £40, or £50, or some other multiple of £10.

But Planck's version of the radiation laws was a little more subtle than this. It said that the 'size' of each piece of electromagnetic energy – each 'quantum of radiation' – radiated by a black body must be proportional to its frequency, obeying the rule $E = hf$, where E is the energy, f is the frequency, and h is a 'new' constant of physics, unsuspected before 1900, which is now called Planck's constant.

It is easy to see how this gets rid of the ultraviolet catastrophe, once the statistics are put in. The picture physicists have of a hot object is that it is made up of a large number of atoms, which can each emit energy. But not all the atoms have exactly the same energy to radiate. The average amount of energy corresponds to the temperature of the black body, and to the peak of the corresponding black-body curve. But some atoms have more energy than the average, and some less, and the numbers of atoms on either side of the peak in the distribution is given by a precise statistical formula. For very high frequencies (short wavelengths), the energy needed to emit one quantum of radiation is very large. But only a few of the atoms have this much energy, and so only a few of these high-energy quanta can ever be emitted. The higher the temperature, the more high-energy atoms there are and the more high-energy quanta can be emitted, so the more the actual spectrum resembles the ultraviolet catastrophe. For very low frequencies (long wavelengths), even if there are many atoms with enough energy to produce the appropriate quanta of radiation, each quantum is so feeble that even adding them all together doesn't produce much energy. Only in the peak of the distribution will there be a lot of atoms each with enough energy to emit moderate-sized lumps of radiation which add together to produce the peak in the spectrum and give the object the colour characteristic of its temperature.

Before Einstein, although many physicists were happy to accept Planck's method of resolving the ultraviolet catastrophe, nobody thought of it as much more than a mathematical trick. There was no suggestion, not even from Planck himself, that light, or other forms of electromagnetic radiation, really did only exist in the form of little lumps, or quanta. But Einstein was different.

You can already see, from this brief résumé of the birth of the quantum theory, that it was a natural step for Einstein, with his growing expertise in statistical matters, to take a look at the implications of Planck's work. It is not the case, as you might get the impression from reading some accounts, that Einstein's three great papers of 1905 bore no relation to each other (or to his thesis). Far from it; everything was interwoven, and the common thread linking all the work we have described so far is the statistical-mechanics approach to thermodynamics.

Einstein actually touched on Planck's work in 1904, in the only paper he (Einstein) published that year.[6] It was in that paper that Einstein first took the bold step of applying statistical techniques to radiation, and this approach carried over into the March 1905 paper best remembered for its explanation of the photoelectric effect, which was actually the very next paper that he published. There was much more in that paper, though, than the Nobel Prize-winning work itself. In particular, there was yet another derivation of Avogadro's number, linking the paper in even more tightly with Einstein's other work of the time.

It may seem surprising that Avogadro's number can be derived from measurements of the properties of radiation, but Planck himself had appreciated this, and Einstein developed his own version of the relevant calculation. The reason why Avogadro's number can be inferred from studies of radiation is precisely that both the radiation and a box of gas obey the same laws of thermodynamics, including the second law, and statistics. There is an expression for entropy, for example, which was originally calculated by Boltzmann from the kinetic theory of gases, and contains a number which is therefore called the gas constant, but which turns up in *any* calculation of entropy on a statistical basis, whether you are dealing with a gas or not. The gas constant can be expressed in terms of Avogadro's number – it is equal to that number multiplied by another constant, usually written as k and known as Boltzmann's constant. So when his statistical calculations of the properties of radiation gave Einstein (and Planck before him, using a slightly different approach) a value of k, he could immediately use this and the known value of the gas constant to work out Avogadro's number. In the March 1905 paper, the value that Einstein came up with was 6.17×10^{23}, astonishingly close to the modern value of 6.02×10^{23}.

So when Einstein chose the round value of 6×10^{23} for the sample calculation in his Brownian motion paper a few weeks later, it wasn't just a lucky guess, but came from his own calculations of, if you like, the thermodynamics of light. It is interesting, though, that Einstein did choose this value of the number, and not the slightly smaller value that he had come up with from the sugar-solution calculations in his thesis. At the time, he did not comment on whether one estimate was better than the other. They were both based on valid science, and they were both in the same range as estimates arrived at by other means. But it is hard not to believe, in the light of everything else we know about Einstein, that he instinctively regarded the value he had derived on the basis of the application of fundamental principles and a relatively simple measure of the spectrum of a black body as being more reliable than a value which depended on the rather cruder behaviour of molecules in solution, and on experiments that were one step further removed from the basic physics. For once, though, it was a known experimental puzzle that led Einstein into his next step, the great philosophical leap of accepting the physical reality of light quanta, or photons, as we would now call them.

Back in the 1880s, some physicists had noticed, during experiments with electricity, that shining ultraviolet light on to a metal surface could make that surface acquire a positive electric charge. In 1899, the British physicist J. J. Thomson (always referred to by his initials) proved that negative electric charge is carried by particles, now called electrons; he later received the Nobel Prize for this work. During the investigations that led to this discovery, Thomson also suggested that the buildup of *positive* charge on a metal surface under the influence of ultraviolet light was a result of *negatively* charged electrons being emitted from the surface. But the key work that led directly to Einstein's 'discovery' of photons was carried out in 1902 by the Hungarian researcher Philip Lenard.

Lenard's experiments established two things about this photoelectric effect. First, provided that the colour of the light shining on the metal is the same, the electrons that are knocked out of the metal by the light all have the same energy, no matter how bright (or how dim) the light is. This is not what you would expect from everyday common sense. A brighter light has more energy, and you would think this would give the electrons a bigger kick, so that they would fly out of the metal with more energy. But that is wrong. For a particular colour of light (which means a particular frequency), if you double the strength of the light you may get twice as many electrons 'boiling off' from the metal each second, but each of those electrons moves with the same speed, which means that they have each absorbed the same amount of energy from the light.

The second discovery Lenard made is that if you change the colour of the light, then you do change the amount of energy carried by each electron, and therefore the speed with which they move. Even for light sources that are all the same brightness, the energy carried by the electrons turns out to be proportional to the frequency. So a red light shining on a metal surface, for example, will cause the emission of electrons all with the same energy, no matter how bright the light, and that energy will be less than the energy of the electrons emitted when a blue light shines on the same metal surface, no matter how dim that second light may be (blue light has a higher frequency, and shorter wavelength, than red light).

Einstein explained this in his March 1905 paper. He did so by suggesting that a light beam really does consist of a stream of quanta (photons) with energy given by Planck's formula $E = hf$. In that case, one electron is knocked out of the metal surface when one photon hits one atom (notice how this explanation interweaves with the work establishing that atoms really do exist). For a particular frequency (colour), every photon will have the same energy. So in every case, the emitted electron will have the same energy. The difference between a bright red light and a dim red light is not that each photon has more energy in the bright light, said Einstein in effect, but that there are simply more photons in the bright light. But since f is bigger for blue light, so is E, and each electron emitted by the photoelectric effect when a blue light shines on the metal will have more energy than the electrons emitted under the influence of the red light.

Einstein's interpretation fitted all the experimental evidence about the photoelectric effect. But by giving the light quantum a physical reality, it seemed to fly in the face of a hundred years' accumulation of evidence that light was a form of wave, and the more recent discovery, linked with Maxwell's equations, that it was a form of electromagnetic wave. For ten years, many physicists found it very difficult to believe that Einstein's explanation of the photoelectric effect could be anything more than another mathematical device, with no physical reality. After all, how could light be *both* particle *and* wave? As we shall see in Chapter Ten, this wave-particle duality in fact lies at the heart of the new quantum physics that was to be established in the 1920s, and that underpins our understanding of the subatomic world today. Maxwell proved that light is a wave; Einstein proved that light is made of photons. Both were right.

Einstein's work on the photoelectric effect was finally proved correct by a superb series of experiments carried out by the American Robert Millikan, culminating in a paper published in 1916. The particular significance of this work is that Millikan started out an avowed opponent of the idea of light quanta, and was trying to prove Einstein wrong. After years of effort, he succeeded in persuading himself that Einstein was indeed right, and in finding a very accurate value for h, Planck's constant, of 6.57×10^{-27}. Looking back a quarter of a century later, Millikan ruefully commented: 'I spent ten years of my life testing that 1905 equation of Einstein's and contrary to all my expectations, I was compelled in 1915 to assert its unambiguous verification in spite of its unreasonableness.'[7]

But one of the most breathtaking aspects of the way Einstein worked in his early years, and especially during the *annus mirabilis*, is the way in which he refused to allow any expectations to cloud his investigations of the nature of the world. Having broken the mould of a hundred years by establishing that light behaved as a stream of particles, within a few weeks he was making another great leap into the unknown by starting out from Maxwell's equations and the well-established nineteenth-century fact that light was a wave. The March 1905 paper suggested the reality of photons. In April, the thesis was completed. In May came the first of Einstein's major papers on Brownian motion. And in June 1905 he sent off to the *Annalen der Physik* the paper which set out the special theory of relativity, resolving the conflict between Maxwell's equations and Newton's laws in Maxwell's favour.[8]

This paper really does stand distinct from the other work of that great year, dealing with the nature of space and time and the dynamics of individual moving bodies, not with the statistical behaviour of large numbers of particles, be they molecules or photons. And yet, even here there is a thread of continuity running through Einstein's work. He had, after all, been thinking intensely about the nature of light earlier in the year, while investigating the photoelectric effect. And he later said that it was by thinking about the nature of light that he first set out on the trail to the special theory of relativity.

The question Einstein puzzled over was this. If you could travel alongside a beam of light, at the same speed as the light, what would you see? Common sense would say that you would not be able to detect any wave motion in the light beam, because you would be riding alongside the wave. The wave would seem to be frozen in space. And yet, if the electric part of the wave were not waving, it would not create the magnetic part of the wave. And if the magnetic part of the wave were not waving, it would not create the electric part. There would be no wave! Something must be wrong with the naive picture, and by the end of June 1905 (his paper was received at the offices of *Annalen* on 30 June) Einstein knew what that something was.

The special theory

Einstein's special theory of relativity was based on two fundamental postulates. First, echoing Newton, that the laws of physics are exactly the same for all observers that move at constant velocity relative to one another. In other words, if we set up a physics laboratory in a spaceship coasting in one direction at, say, half the speed of light relative to a certain star, and you set up a physics lab in a spaceship coasting at half the speed of light in the opposite direction relative to that same star, then when we carry out identical experiments in our two laboratories we will get identical results. This is equivalent to saying that all such observers are entitled to say that they are at rest, while any other inertial observers are moving, and there is *no* special inertial frame. The second postulate of the special theory takes Maxwell's equations at face value, and says that for such observers (in what are called 'inertial frames'), any experiment to measure the speed of light will give the same answer (usually written as c, since it is a constant), regardless of how the source of the light is moving relative to the observer.

In the example we have just given, in *both* laboratories measurements will show that the speed of light coming from the chosen star (and, indeed, from any other star) is c, *and* that the speed of light coming from a lamp in the laboratory is *c*, ***and*** that the speed of light signals being flashed from one spaceship to the other is *c*.

It is this constancy of the speed of light that leads to all the non-commonsensical predictions of Einstein's theory, that a moving ruler shrinks and gets heavier, while a moving clock runs slow. It also says that the way to add up velocities is ***not*** the way Newton believed. If, for example, an observer sees one spaceship fly past him at 0.75*c* (three-quarters the speed of light) and another spaceship fly by in the opposite direction, also at 0.75*c*, then observers in either spaceship will measure the relative speed of the other spaceship not as 1.5*c*, but as 0.96*c*.

Einstein actually arrived at the equations that make these predictions by thinking about how a pulse of electromagnetic radiation spreading out spherically from a light source would look to observers moving at different velocities. The sphere that is filled by the light after a certain time has a radius equal to the speed of light multiplied by the time that has elapsed (*ct*). But it can look spherical to all moving observers only if their measuring rods are shrunk by their motion relative to the point the light is spreading out from. And the sphere ***must*** look spherical to all inertial observers if Einstein's first postulate holds. Because the size of that sphere after a certain time is simply *c* multiplied by the elapsed time (*t*), the expression *ct* comes into the equations, exactly like a measure of length, but with the opposite sign ('−' instead of '+'), so that time can be regarded, in the special theory of relativity, as a kind of negative length.

Now, we don't intend to take you through Einstein's reasoning here, because a few years after the June 1905 paper was published, Hermann Minkowski, one of Einstein's old teachers, came up with a simple and elegant picture – literally, a picture – of what it all means. Minkowski said that time could be literally regarded as a fourth dimension, like the three familiar length dimensions of space ('up/down', 'left/right' and 'forward/backward'), but multiplied by *c* and with a negative sign in front of it. In a lecture given in Cologne in 1908, he said:

> The views of space and time which I wish to lay before you have sprung from the soil of experimental physics, and therein lies their strength. They are radical. Henceforth space by itself, and time by itself, are doomed to fade away into mere shadows, and only a kind of union of the two will preserve an independent reality.[9]

This union of space with time is now known as spacetime. Since there are three dimensions of space, and one dimension of time, spacetime has four dimensions. What Minkowski did in 1908 was to explain Einstein's special theory of relativity in terms of the geometry of four-dimensional spacetime.

Most people have trouble picturing four dimensions, but don't worry. You can understand what Minkowski meant, and how special relativity works, by looking at a pencil, and the shadow that it casts on your table. The pencil has a definite length in the three-dimensional world, which you can measure. But if you look at the shadow of the pencil on the two-dimensional tabletop, and twist the pencil about in the air, you will notice a curious thing. Depending on the angle you hold the pencil at, the length of the shadow may be zero, or it may be as long as the pencil itself, or it may have any length in between. The length of the shadow in two dimensions depends on the orientation of the pencil in three dimensions.

The special theory, in the geometric picture developed by Minkowski, says that any object, such as a pencil, has a real *four*-dimensional 'length', which is usually called its 'extension', not in space but in space*time*. But depending on how the object is oriented in four dimensions (in spacetime), its actual length measured in our three-dimensional world may have any value from zero up to a length equal to the extension. 'Twisting an object into a new orientation' in four-dimensional spacetime is the same as moving the object at a constant velocity through three-dimensional space. The faster the object moves, the shorter its three-dimensional 'shadow' gets, and the shadow disappears entirely (the length becomes zero) when the object is moving at the speed of light.

Similarly, time in our world is a shadow of an extension in

four-dimensional spacetime. But because of the minus sign in front of the *ct* in the equations, the faster a clock moves, the more spread out its 'shadow' becomes, so that time runs more slowly, and if the clock could move at the speed of light, then time, as measured by the clock, would stand still.

None of these effects show up in any significant way until an object is moving, relative to the observer, at a sizeable fraction of the speed of light. Since the speed of light is so large – 300 million metres a second – the effects never show up in everyday life, which is why they do not seem to be common sense. One second in the time dimension is equivalent to 300 million metres (actually, *minus* 300 million metres) in any of the three length dimensions, so space and time only have an exactly equal footing when you are moving at 300 million metres a second – at the speed of light – and their equivalence only begins to show up for things moving at a respectably large fraction of that speed. But it is important to appreciate that every single one of the predictions of the special theory of relativity has been tested by experiments, and that it has passed every test with flying colours.

The simplest way to see this is in terms of the behaviour of particles called muons, which are created high in the atmosphere of the Earth by the impact of cosmic rays from space. Muons move at a speed very close to the speed of light. But they are also very short-lived particles, and soon (in a couple of microseconds) decay into other kinds of particle. By measuring the number of muons found at high altitudes, using instruments on an uncrewed balloon, and comparing this number with the number of muons arriving at the ground, physicists find (the experiment was first carried out in 1941) that although the lifetime of a muon is too short, according to clocks here on Earth, for them to get through to the ground, in fact most of them do penetrate the atmosphere. The explanation is that time runs slower for the muons, because they are travelling at nearly the speed of light, and therefore they do have time to reach the ground before they decay. At the speed they travel, Einstein's special theory says that time is slowed by a factor of 9.

But how do things look to an observer riding with the muons? Remember that according to Einstein's first postulate they are entitled to regard themselves as at rest, with the Earth rushing past at nearly the speed of light. So their clocks will be unaffected by motion. Instead, such an observer will see the Earth, and the atmosphere, enormously shrunk in the direction of motion, so that the atmosphere is very thin. So thin, in fact, that even though the lifetime of a muon is very short it has time to penetrate from the top of the atmosphere to the ground before it decays. At the speed involved, Einstein's special theory says that the distance is shrunk by a factor of 9.

Either way, a factor of 9 is exactly what you need to reconcile the observations of muons high in the atmosphere from balloons with those made on the ground.

This experiment really has been carried out. It confirms that moving clocks run slow, that moving objects contract, ***and*** that any observers moving at constant velocity are entitled to regard themselves as at rest.

There are many more experimental confirmations of the special theory, but we will mention just one. It turns out, from Einstein's equations, that an object travelling at less than the speed of light can never be given enough energy to make it accelerate to the speed of light (let alone to go any faster). You never can run alongside a beam of light at the same speed, and so the puzzle of how the waves would look if you rode with them does not arise. Instead, the more energy you put into a moving object, the faster it will go, but the extra speed it picks up is always less than the amount needed to take it past the speed of light. As you put more and more energy into a moving object, such as the particles that physicists whiz around in accelerators like the ones at CERN, in Geneva, the less each extra bit of energy increases the speed of the object. You get less return (in the form of extra speed) for the same investment (in the form of extra energy). But the extra energy must go somewhere, and the equations tell us that it goes into the mass of the particle, which goes a little bit faster but gets a lot heavier as it continues to accelerate and gets near the speed of light.

Again, this effect has been measured, directly, in countless experiments at particle accelerators around the world.

*

Minkowski's comment that the strength of the special theory of relativity derives from its roots 'in the soil of experimental physics' gets more apposite with each decade that passes. And the equivalence between energy and mass implied by this effect is what gives us the most famous equation in all of science, $E = mc^2$. It says that mass and energy are interchangeable, and that just as energy can, under the right circumstances, be turned into mass, so mass can, under the right circumstances, be turned into energy.[10] The most famous example of that equation at work, of course, is in a nuclear bomb, where a tiny amount of mass is converted (thanks to the multiplying factor c^2, which works out at 90 million billion, in metric units) into a great deal of energy. There is indeed no doubt that the special theory of relativity is an accurate description of reality.

Before we leave the special theory, though, we want to clear up one puzzle which worries very many people when they first meet it. It is sometimes called the 'twins paradox', although it isn't really a paradox at all.

It goes like this. Suppose you have a twin sister, exactly the same age as yourself. While you stay at home, here on Earth, your intrepid sister flies off in a spaceship, at a speed that is a sizeable fraction of the speed of light. After a while, she turns the spaceship around, and flies back. During the flight, her clocks must run slow, relative to your clocks here on Earth. She must, according to Einstein's theory, age less than you have. But if she can say that she was at rest, and you were the one doing the moving, she will say that your clocks were running slow. You predict that your sister has aged less on her journey than if she stayed at home, while she says you must have aged less. Each twin predicts that the other one is younger!

The resolution of the 'paradox' lies in the fact that your twin had to turn her ship around in order to come home. She changed the direction she was moving in, not just through space but through spacetime. If she had kept on travelling in a straight line through space at a constant speed, she would have been travelling in a straight line through spacetime and both of you would indeed have been entitled to argue that the other twin was the younger one. But there would have been no paradox, since under those circumstances you could never again stand side by side and compare notes.

By coming back home, however, your twin has not kept to a straight-line journey through spacetime. Your own 'journey' is indeed a straight line, because you always travel at the same velocity, but in order to get back to you she has to travel around two sides of a triangle in four-dimensional spacetime.

In the familiar three-dimensional world, the distance from one point to another in a straight line is always less than the distance between the same two points along the other two sides of a triangle. The same is true of the distance part of your sister's journey – she has indeed travelled more kilometres than you have. But this is not the whole story. Because of that minus sign in front of the *ct* in the equations of special relativity, the time difference between two points in space-time is always *less* if you go around the two sides of the triangle, instead of following the straightest possible line from one point to the other. The extra distance travelled by your sister is exactly balanced by the smaller time her journey takes, so that you and she both 'use' the same amount of spacetime and both start out and end up in the same place at the same time. So your twin sister who has been on the journey really is younger than you are when she comes home.

It may not be common sense, but at least you can see that there is a difference between the two journeys through spacetime, one in a straight line and the other round two sides of a triangle. And, again, this prediction of the special theory has actually been tested, using extremely accurate timepieces called atomic clocks, flown around the world on jet airliners and compared with identical stay-at-home clocks in their return to the laboratory (such tests were carried out for the first time in 1971). Even though the effect is tiny for journeys at such low speeds, compared with *c*, the clocks are accurate enough to measure the difference, which exactly matches the predictions of relativity theory.

The special theory of relativity is a proven, tried and tested theory of the way the world works. Einstein himself did not, at first, welcome Minkowski's geometrisation of his theory, but it was Minkowski's way of presenting the theory that led to its widespread acceptance among physicists.[11] As early as 1912, Einstein was nominated for the Nobel Prize for this work, and although he did not receive the prize that year (or ever, for the special theory of relativity), this indicates the high esteem in which he was held at the time. No longer a humble Patent Office clerk who wrote physics papers in his spare time, Herr Professor Einstein was a fully fledged member of the European academic community, with an established reputation. And yet, his greatest work was still to come.

Chapter Seven

The Peripatetic Professor

Einstein's move into academia in the autumn of 1909 was long overdue, but it did not change his domestic life substantially. Although, a year after his appointment, his salary was increased to 5,500 Swiss francs per annum, this did not in fact go as far as 4,500 francs had done at the Patent Office, for now he was required to maintain a certain image befitting the social status of a university professor. In his new role he and Mileva were expected to host dinner parties and functions at their home. He had to own formal dress for university functions and robes for ceremonial occasions. All this cost money. As a consequence, it was decided that, once settled into their new flat on Moussonstrasse, the Einsteins would take in student lodgers to augment the family budget.

There is no doubt that Einstein was delighted by his appointment. Being a member of a university faculty would open many doors to him and ease the path of his scientific labours. It gave him greater resources and far better communication with the rest of the scientific community. However, the appointment at Zurich was as an associate professor or professor 'extraordinary'. With the position came a significant number of teaching hours per week, and although Einstein did enjoy teaching, it distracted him from his research. Soon after taking the new position, therefore, he was keen to move up the career ladder in order to liberate more time for his own research.

Einstein's teaching responsibilities in his new job entailed the supervision of a small group of PhD students and six to eight hours of lectures and seminars per week. This does not sound too demanding but, because of his limited teaching experience and his general approach to the task, preparation was a drain on his time. In 1909 he wrote to his friend Michelangelo Besso: 'I am *very* occupied with the courses, so that my *real* time is less than in Bern.'[1]

Einstein was a very popular lecturer and he had an easy rapport with his students. He was never a man to adopt superior airs and, unlike many other members of the faculty, he was never too busy to sort out the problems of individual students. His students saw Einstein as eminently approachable and the usual formalities which separate lecturer and student simply did not exist with him. He socialised with some members of his classes and professor and students were often to be seen together discussing physics in the cafés of the city. One of their favourite haunts was the Terrasse Café in the centre of the city. If this did not suit the occasion, a small group with Einstein at their head would often arrive unannounced at the flat on Moussonstrasse and sit around the dining room drinking coffee and continuing with their discussions.

The period spanning his arrival and first year's teaching in Zurich marks a time when Einstein was comparatively quiet in his work on the general theory of relativity. The overwhelming reason for this was the fact that he was involved in other scientific work as well as his demanding responsibilities at the university.

On 28 July 1910, Mileva gave birth to the Einsteins' second son, Eduard. The pay rise of that year probably helped financial matters a little but Einstein must have felt, not for the first time, the irksome problems presented by the need to maintain a reasonable standard of living while trying to push ahead with his *raison d'être* – his scientific work.

The family were happy in their new environment. The serious differences between Einstein and Mileva which were later to split them apart were still some years in the future. The sad and painful life of Eduard, who was to die in a Zurich mental institution some fifty-five years later, had only just begun. In the summer sunshine of this city beloved by the young couple, life must have seemed good despite the pressures of domestic responsibilities. In comparison to the dark years of World War I, the traumas of divorce and, later, the terrors of Nazism, the Einsteins' first decade of marriage now seems an oasis of tranquillity, a time of peace in which Einstein was to find his academic feet and lay the foundations for future greatness.

That first lectureship in Zurich was to be short-lived. Anton Lampa, head of the physics faculty at the University of Prague, approached Einstein towards the end of 1909, at the end of his first term in Zurich, with a view to appointing him to the newly vacant chair of theoretical physics at the German University of Prague.

Lampa was sounding out Einstein's willingness to move, though the search committee in Prague was not to convene until January 1910. Lampa wanted Einstein for the position and as chairman of the committee he was in a very strong position to make his wishes reality. However, Einstein was asked to keep the whole matter secret for there were other candidates and other opinions to consider. As Einstein was to discover during the course of the next year, by saying this, Lampa was not exaggerating.

The other major candidate for the appointment at Prague was the physicist Gustav Jaumann, a professor at the Technical Institute in Brno. By all accounts, Jaumann was an arrogant and vain man who held dogmatically to the views of Ernst Mach, the father of logical positivism and a man whose own scientific ideas were often at variance with Einstein's. Mach did not agree with much of what Einstein postulated in relativity.

Einstein's work was of great and growing importance and his publications were far more impressive than Jaumann's. Max Planck had written to the faculty committee at Prague, saying: 'If Einstein's theory should prove to be correct, as I expect it will, he will be considered the Copernicus of the twentieth century.'[2] Nonetheless, the committee voted against Lampa and appointed Jaumann to the post.

The fundamental reason for this unexpected turn of events was the matter of nationality. Jaumann was an Austrian and Prague was part of the Austrian Empire. Einstein was Swiss, and German by birth. The committee, through the Austrian Ministry of Education and then on up to the emperor, had decided that they wanted an Austrian physicist for the position even if his abilities were plainly not in the same league as Einstein's.

Once again, Einstein had been competing with an unusual candidate. When he discovered what had happened, Jaumann's ego could not allow him to accept the post and he handed it over to Einstein, declaring to the committee: 'If Einstein has been proposed as first choice because of the belief that he has greater achievements to his credit, then I will have nothing to do with a university that chases after modernity and does not appreciate true merit.'[3]

This was not to be the end of the matter. The Austrian emperor, the elderly Franz Josef, had decreed that professorial positions at universities in the empire should be granted only to members of an officially recognised religion. Einstein had rejected his Jewish background while still a teenager; however, he had never done so officially. Now he had either to declare his agnosticism or to play the game and put down something to satisfy the authorities. He decided upon the latter choice and announced his faith as 'Mosaic', the Austrian name for the Jewish faith. This satisfied the emperor and the appointment was finally agreed upon. The whole process had taken over a year and it was not until April 1911 that the appointment was made official.

The move to Prague was not at all popular with Mileva, and Einstein's time there was not altogether happy. Both he and Mileva never overcame a feeling of being outsiders in a foreign land and they never quite managed to settle comfortably into their new environment. In retrospect it is clear that this period in Austria was the point at which cracks began to appear in the Einsteins' marital relationship and that the decisions Einstein made in the next few years, particularly with respect to academic appointments, were to navigate their marriage on to the rocks of divorce.

The Prague of 1911 was a city steeped in history and seething with political undercurrents. Within a few short years of Einstein's arrival there in March 1911, the city would be at the epicentre of European upheaval. Little more than seven years after the crisp spring day on which Albert Einstein took up his new appointment, the Habsburg Empire of which Prague was a part would be no more and the city itself would become the capital of the newly created state of Czechoslovakia.

In many ways, Prague was a divided city and a microcosm of the broader European arena of the early part of the twentieth century. The most important division in Prague was that between Czech and German. The two peoples very rarely interacted – a fact evident in all aspects of life. They lived in different parts of the city and there were two sets of theatres, concert halls and other cultural and social institutions, and two separate universities, the German and the Czech University.

Although only around 5 per cent of the population of Prague were German-speaking in 1911, they exuded a general air of superiority in the city, which inflamed Czech national sensibilities. In taking up his position at the German University, Einstein was obliged, much against his inner feelings, to accept Austro-Hungarian citizenship. He was, however, allowed to retain his Swiss papers and thus for this period had dual nationality. Further complications arose because of the fact that he was Jewish.

Although approximately half the Germans in Prague were Jewish, Jews were placed in a peculiar position. On the one hand the Jewish community was distrusted by the Czechs as 'spies' and 'German sympathisers', and on the other they were experiencing a growing hatred from a section of the non-Jewish German population, a hatred which within a generation would develop into the mentality behind the death camps.

It is quite likely that Einstein was aware of these troubles before he arrived in Prague; he certainly had plenty of time to think through the consequences of the move. Nonetheless, he still decided to go.

On the plus side was the fact that the University of Prague had appointed him a full professor. With the increased prestige came a larger salary and the university could offer the scientist far greater resources than he had at his disposal in Zurich. The Einsteins' standard of living certainly improved upon moving to Prague. They were able to employ a maid after the move, and having to take in student lodgers was now a thing of the past. Their Prague flat boasted electric lighting rather than the gas system they had had in their previous homes.

Apart from the improved resources at the university, Prague was a great cultural centre in its own right. During the short time Einstein lived in the city, he made friends with a number of important figures within the Jewish intellectual society and outside the realms of science.

He met Hugo Bergmann, a Zionist campaigner, but was not then inspired to take up the cause. Einstein's Zionist leanings and his bitter-sweet relationship with that cause were still some years away.

Einstein also knew Franz Kafka. At the time the twenty-eight-year old writer, who had obtained a doctorate at the university in which Einstein now worked, was just beginning to find his way in his own paranoid fictional world, while holding down a job as a clerk at the Workers' Accident Insurance Company. He wrote 'Metamorphosis' at exactly the time he would have known Einstein. The two men were certainly friends but did not have a close relationship. As Einstein had done only a couple of years earlier, Kafka led a dual life, working in a mundane boring job to pay the bills and in his spare time fulfilling his true destiny, writing what would be appreciated by future generations as great works of literature.

Another member of the same circle was the novelist Max Brod, now best known as the editor of Kafka's papers. He was to write a novel called *The Redemption of Tycho Brahe*, about the sixteenth-century astronomer who had spent the last two years of his life in Prague, where his assistant, the great Johannes Kepler, lived and continued to work after Tycho's death. Brod is said to have based his characterisation of Kepler on the young Einstein he knew for a short period in Prague, or so Einstein was told by his friend Walther Nernst, who had read the novel.

Although Einstein enjoyed the company of nonscientists throughout his life and appreciated a diversity of conversation outside the realms of physics, his closest friend during his time in Prague was a mathematician named Georg Pick.

It was Pick who helped to guide Einstein in the right direction for mathematical assistance with the general theory. Mathematics was often to be Einstein's weakness when it came to formalising his theories, and he was always fortunate to meet others who could help him in this way. Pick led Einstein to the work of the Italian mathematicians C. G. Ricci and Tullio Levi-Civita, which was to prove instrumental in the development of Einstein's ideas during the next few years. Pick was also a keen musician and introduced Einstein to a circle of musicians in Prague. Einstein joined a quartet which met regularly to play in each other's houses during the evenings.

Although there were many happy aspects to Einstein's life in Prague, the political atmosphere was uncomfortable and after a few months he was already beginning to feel oppressed by the air of formality which surrounded his position at the university. Soon after his arrival in the city, Einstein wrote to his friend Besso: 'My position and my institute give me much joy, only the people are so alien to me.'[4]

As well as being Einstein's first year as a full professor at Prague, 1911 was a year in which Einstein met, in many cases for the first time, all the great names of the scientific community. The occasion for these encounters was the First Solvay Congress held in Brussels in the autumn.

The Belgian industrial chemist Ernest Solvay, the inventor of a process to make the commercially important compound sodium carbonate, had been persuaded to put up the money for the congress by his friend Walther Nernst. Nernst knew that Solvay had his own, somewhat unorthodox theories in the field of nuclear physics. Although Solvay was highly successful in his own field, his theorising in physics was quite off beam. However, Nernst, with an eye to the main chance, had suggested that Solvay finance a great meeting of the leading physicists of Europe to discuss the latest developments in the subject. This would also create a setting for the presentation of the chemist's personal theories.

This Solvay agreed to and Einstein, along with the cream of the physics community of Europe, arrived in Brussels at the end of October 1911 for a five-day conference. Solvay delivered a talk in which he outlined his theories and the gathered scientists politely sidestepped giving their opinions about the host's exotic ideas. Each of the invited physicists gave a lecture and a good deal of time was put aside for discussion and communication of new developments.

The list of twenty physicists present at what was to be the first of many Solvay congresses reads like a who's who of physics in the first quarter of the twentieth century. The guest list included Marie Curie, Max Planck, Nernst, Ernest Rutherford, Jules Henri Poincaré, Hendrik Antoon Lorentz, Maurice de Broglie and of course Albert Einstien, at thirty-two one of the youngest present.

The Solvay Congress of 1911 and the others which followed were enormously successful in bringing together the great names and important figures in the world of physics. Its instigation was an encouraging move against the current trend towards disunity in Europe and saw the coming-together of the leading personalities who were to shape the future of physics across national barriers and politically contrived lines of demarcation.

The Solvay Congress also did a great deal for Einstein's image and status in the scientific community. Making friends with many of the other scientists at the congress was undoubtedly helpful in shaping his future career.

In the year leading up to the congress, Einstein had been approached by a number of institutions. The most significant offer had come from the University of Utrecht and for a short time he was tempted to accept their offer. The Dutch physicist Hendrik Lorentz, who was professor of theoretical physics at Leiden, where Einstein had delivered an important lecture in January 1911, was very keen to have Einstein in the Netherlands and used all his persuasive powers to entice his colleague away from Prague.

Einstein and Lorentz had got on very well and the Einsteins had stayed at the Lorentz family home during the lecture visit there. Lorentz was more than aware of Einstein's abilities and was at the forefront of those who believed that the young physicist had yet to deliver his greatest work. Lorentz was of the profound belief that Einstein's contribution, already quite staggering for any one scientist, would prove to be of exceptional importance to the future of physics.

However, despite his best efforts, Lorentz failed. Einstein was not to be enticed to Utrecht, but was instead drawn away from the unsatisfactory situation in Prague by a totally unexpected return to Zurich. The offer came from none other than the institution which had first rejected him as a student and then, because of the feud with Weber, as a teacher – the Eidgenössische Technische Hochschule, or ETH.

By 1911, Einstein had become highly sought after in the world of science. He was still totally unknown to a wider general public, but among physicists he was seen as a leading light. This much can be seen from the glowing references given him by two of the continent's most eminent scientists, Madame Curie and Jules Poincaré, before his appointment to the ETH. In her letter to the selection board, Madame Curie said of her German colleague: 'I much admire the work which M. Einstein has published on matters concerning modern theoretical physics. I think, moreover, that mathematical physicists are at one in considering his work as being in the first rank.'[5] And Poincaré made the following recommendation: 'The future will show, more and more, the worth of Einstein, and the university which is able to capture this young master is certain of gaining much honour from the operation.'[6]

If there were any doubts about Einstein's suitability for the position at the ETH, these references must have swept them aside. So, less than eighteen months after arriving in Prague, the Einsteins were moving back to Zurich.

The ETH was a more important institution than the university. The former was a federal organisation, whereas the university was a cantonal institution governed by a local authority. Therefore, Einstein's move from the university to the ETH via Prague was actually an upward journey. Furthermore, Einstein must have experienced an element of self-satisfaction in acquiring a professorship at the college which had twice rejected him in his youth. In the late summer of 1912, when Einstein took up his new position at the ETH, he had finally reached a stable, financially satisfactory position in the academic world. Perhaps he expected to spend the rest of his career in Switzerland, the country he loved and where he always felt most at ease.

Mileva was pleased by the move. She had always felt uncomfortable in Prague and missed Zurich. However, from correspondence, especially with his friend Besso, it is clear that by 1912, Einstein's marriage was not in the best of shape and there remained very little common ground between the couple. Mileva had lost any scientific curiosity she had once had and showed almost no interest in her husband's achievements despite his growing fame. Perhaps she in some way resented Einstein's new eminence; the fact that she too had received a scientific training but had ended up as a *hausfrau* may have widened the fissure between man and wife. The two boys – '*die Bärchen*', the little bears, as they were known in the family – were demanding on Mileva's energies, and although Einstein loved his children, he was never much of a family man. Their children were not an element in bonding the couple together.

Einstein's stay in Zurich turned out to be not much longer than his period in Prague. Soon after his return, he was made an offer he could not refuse.

Despite the brevity of his stay, Einstein's time in Zurich was highly productive. During the upheaval of frequent moves around central Europe, he had been developing the general theory of relativity. Back once again in Zurich, he teamed up with his old friend Marcel Grossman, who by this time had become professor of mathematics at the ETH. Together they began to form the theory into a rigorous mathematical structure. This led to the publication of a partially successful version of the general theory in 1913, which was later refined and improved, until it had evolved into the fully fledged theory of 1915.

As had been the case at the University of Zurich a few years earlier, Einstein had barely completed a term in his new position when he was approached by representatives of a foreign university endeavouring to acquire his services. He had not been short of other offers during this period. Even before his move back to Zurich, at least one American institution had offered him a lecture tour and there were an increasing number of universities around Europe who were head-hunting him – a definite sign that news of Einstein's genius was spreading.

In the event, it was a couple of close colleagues in the physics community who managed to convince him to make yet another move so soon after settling in Zurich. These two were the Germans Walther Nernst and Max Planck.

In retrospect it is easy to see the arrival of Nernst and Planck in Zurich in the spring of 1913 as a picturesque cameo stage-managed against the backdrop of total chaos about to engulf the whole of Europe little more than eighteen months later. But even if we take the occasion out of context, it is clear that the University of Berlin chose their two representatives with great care, and that Nernst and Planck were most definitely men with a mission.

They made an interesting couple. Einstein knew both of them very well from conferences and correspondence. Planck, for whom Einstein had the greatest respect, was one of the great figures in the development of the quantum theory. He was a severe-looking man, an intensely patriotic German, for whom service to his country was both an honour and a duty. In comparison, Nernst was a breath of fresh air – a gregarious and friendly, commercially minded man who had coupled the creation of a successful business with a career as a physical chemist of genius. He was one of a small but powerful group of entrepreneurial European scientists of the era, a club which had also included such luminaries as Alfred Nobel and Ernest Solvay.

The deal offered by Planck and Nernst was a three-part package. Firstly, it included a professorship at the University of Berlin, a position specially created for Einstein and entailing as little teaching as he wished to undertake. Planck and Nernst knew that this would be an important factor in any offer made to Einstein. They understood the need for time to concentrate on research and that teaching, even for someone who enjoyed it, was an encumbrance. Secondly, Einstein was offered membership of the highly prestigious Prussian Academy of Sciences, Germany's oldest and most important scientific institution, an elite club equivalent to the Royal Society in

London – membership by invitation only. The Academy seat came with a salary, a privilege not extended to all members. Thirdly, he was offered the directorship of a new physics institute.

Together, the ingredients must have seemed irresistible. There would be enough money involved to release Einstein from all non-research commitments and the best facilities and resources available anywhere at the time. The directorship of the proposed institute would not, he was assured, distract him too much from his primary role. The Germans wanted Einstein for his scientific work, not as a teacher or an administrator. Even so, it appears that he had to ponder the matter for a while. Discussions had actually begun in 1912 and Einstein had made at least one visit to Berlin in that year with the sole intention of discussing the proposal. By the spring visit of Nernst and Planck in 1913, they were keen to come away with a definite answer one way or the other.

After lengthy talks in Einstein's rooms at the ETH, he was still unable to give a definite answer to the two men. He suggested that they visit the Zurich landmark, the Rigi, which involved a journey by tram and cable car, while Einstein sat and chewed over the matter. He promised them that by the time they returned he would have an answer. He would meet them at the station, he said, and they would know immediately whether or not he had accepted their offer; if he was carrying a red rose, he would take the job; a white rose would mean that he was turning them down.

The two men went to admire the view from the Rigi, no doubt amused by their colleague's eccentricities and anxious about the outcome. They need not have worried. Upon their return, Einstein was waiting for them on the platform holding a red rose.

By the summer of 1913, there were still a number of uncertainties and confusions about the appointment which needed to be ironed out. Most significant of these was the offer of the directorship of the Kaiser Wilhelm Institute which at the time of Nernst and Planck's visit had not actually been established.

The idea for a collection of institutes for advanced research had originated some years earlier when Kaiser Wilhelm II had been convinced by leading scientists and industrialists that high-quality research conducted in a country was not only a status symbol but also a great source of industrial and economic development. From scientific research came the ideas that created wealth in future decades. Germany had at the time already gained considerable benefits by developing its scientific know-how, and had the most advanced and profitable chemical industry in the world.

The Kaiser was greatly influenced in this area by his permanent secretary in the Ministry of Education, Friedrich Althoff. Realising that men who had made a great deal of money in industry often also craved a certain respect in intellectual circles – to compensate, perhaps, for what many saw as the uncouth nature of commerce – Althoff decided to exploit this weakness. He suggested that the Kaiser invite leading industrialists to dinner and, over dessert and a fine wine, propose the idea that, in return for a financial contribution, they would be made 'senators' of an academic research establishment. With the honour would come the absolute necessity of wearing elaborate gowns on ceremonial occasions and, of course, contributors would have their names attached to a worthy and highbrow intellectual establishment.

The plan worked and the funding was arranged for the beginning of the programme. By 1917, three years after Einstein had accepted the appointment to Berlin, the Kaiser Wilhelm Gesellschaft was opened under his directorship.

Another hurdle was again the question of nationality. It is easy to see that despite the wonderful opportunities offered to Einstein at this point in his career, he would feel very strongly that he did not want to regain Prussian citizenship after fighting so hard to discard it. He made it a condition of the deal that he would be able to retain his Swiss nationality, and he was promised this by Nernst and Planck as early as the meeting in 1913, long before the finer details had been settled. It has been claimed however that Einstein regained Prussian citizenship as a consequence of joining the Prussian Academy of Sciences and that it was automatically imposed. Still, for the duration of his stay in Berlin, Einstein believed that he was a Swiss and not a German.

When informed of his decision to accept the offer from Berlin, Mileva Einstein was angered and upset. With cynical hindsight it is possible to suppose that Einstein had anticipated her reaction and did not actually see it as a negative point. Aside from all the career advantages the move offered, he may have seen it as an opportunity to precipitate the break-up of his marriage.

Whatever Einstein's motives, the offer was formally approved by the emperor on 12 November 1913 and Einstein accepted the position on 7 December, agreeing to take up his new position the following April. So, the die was cast. Einstein, in a move which was to establish him as one of the leading academics on the Continent, also brought about the events which would mean the death-knell to his eleven-year marriage, separate him from his children and set him on a new course both in his personal life and his career in a country which, as a headstrong and sensitive youth, he had forcefully denounced.

Despite the calamities and traumas which this move would bring, Berlin was where Einstein finally began to settle down. It is ironic that the nation in which he spent the prime of his life and in which he produced his most important work was the very same country for which as a youth he had shown only contempt. Einstein's Berlin period was a time when he was to re-evaluate his world-view, his opinions about war, peace, nationalism and the human situation. It was also the time when Albert Einstein became world famous.

The first move along the road to universal recognition – experimental proof of the general theory – was delayed for five years by the calamity of World War I. The first experiment to attempt to verify Einstein's work was actually proposed as early as 1914. But, despite every effort, this prewar experiment did not take place. As luck would have it, this worked in Einstein's favour, because at the time, his original theory was not fully developed and the experiment would only have shown it to be incorrect. It was to be another five years before experimental verification was achieved, by which time Einstein had greatly improved the general theory.

The experiment to verify the general theory was first proposed by the astronomer Erwin Finlay-Freundlich. Freundlich had encountered the theory in 1911 when a colleague of Einstein's at the University of Prague visited the Berlin Observatory where Freundlich was working. Having gained an early fascination with the implications of Einstein's theories, Freundlich wrote to Einstein with his proposal for proving the general theory by astronomical experiments.

The details of the experimental proof of the general theory is explained in Chapter Eight. It required observation of starlight during a solar eclipse. Freundlich informed Einstein that the next observable solar eclipse was to be in southern Russia in the summer of 1914.

During the course of the next three years, the two scientists developed a close working relationship. This grew into a personal friendship when Einstein moved to Berlin in the spring before the eclipse. Einstein also helped to fund the project. Aside from scientific curiosity, he of course realised the importance of Freundlich's proposed trip. If he were to prove his general theory, it would bring its creator worldwide fame, not to mention the Nobel Prize for Physics.

To obtain the necessary funds, Einstein enlisted the help of his colleagues. Planck was very helpful and had a great deal of influence in powerful circles. Money also came from such unexpected sources as the eminent chemist Emil Fischer. German science was rich and there were a significant number of wealthy scientists around who had made their money from industrial applications of their ideas. Some of them doubtless felt that it was their turn to put something back into the kitty.

By the time Einstein began working at the University of Berlin, the plans for the expedition had been finalised, the money raised and the equipment was tested and ready. The team, headed by Freundlich, set off for the Crimea in the early summer of 1914. Einstein settled into his new post as professor of physics at the university, member of the Prussian Academy of Science and director-in-waiting at the yet-to-be-established Kaiser Wilhelm

Institute. Meanwhile, both in his personal world and in the broader scheme of European politics, things were reaching a new crisis point.

Very soon after making the move to Berlin with her husband, Mileva Einstein decided to return to Zurich with the two boys.

Initially, the return to Zurich was seen as a temporary move. In view of the worsening political climate in the summer of 1914, the family's return to Switzerland was undoubtedly considered a prudent step. However, it soon became clear to all that in fact the Einsteins had separated. Einstein was to live out the major part of the war alone in a bachelor apartment in Berlin while Mileva stayed with the children in Switzerland.

The outbreak of World War I came to many as a great surprise; but the political sores which eventually erupted into four years of bloody conflict had been festering for some time. In the first decade of the twentieth century, Europe had been a quagmire of international hostility; yet, when the conflict actually began, it happened quickly and with horrifying violence.

It is generally considered that World War I began with the assassination of the heir to the Austro-Hungarian throne, Archduke Ferdinand, by the nineteen-year-old fanatic Gavrilo Princip in Sarajevo. However, the mounting political problems in Europe were supported by an unprecedented escalation in the military forces of the leading nations of the continent. The assassination of the archduke gave the Austro-Hungarian Empire the excuse it needed to propel Europe from political and military stalemate to open aggression. Within a few weeks of the shooting in Sarajevo, Britain, France and Russia were aligned against the German and Austro-Hungarian empires. The first casualties of war were the dead and wounded from the carnage of Mons in Belgium.

In Berlin, the anti-militarist Einstein was horrified by the unfolding events. Meanwhile, in the Crimea, instead of bringing the verification of Einstein's theory, Freundlich's expedition had been arrested and imprisoned in Odessa.

Einstein feared that his colleague would be spending the rest of the war in a Russian jail. However, Freundlich and the others were released within a month, having been exchanged for a group of Russian officers, and returned empty-handed to Berlin by September. There Freundlich lived out the rest of the war, assisting Einstein and continuing his work at the Berlin Observatory.

The outbreak of hostilities did not at first affect Einstein's daily life. The general feeling was that the war would be short. It was not only Field Marshal Sir John French, commander of the British Expeditionary Force, who was of the opinion that 'the war would be over by Christmas'. In the summer of 1914, nobody could have foretold the true scale and horror of the next four years.

Sitting in his Berlin office, Einstein could not have imagined what lay ahead for the warring nations of Europe any more than the next man, but when it came to the question of the morality and logic of the conflict he was completely at variance with the vast majority of those around him.

What appears to have initially most shocked Einstein was what he saw as the totally insupportable attitudes of his colleagues and co-workers.

The majority of scientists on both sides supported the war effort of their respective nations. Einstein was wholly against this. He was appalled by the fact that the men for whom he had the greatest respect as intellectuals and thinkers should be taken in by the stupidity of hostile governments. He found it almost impossible to reconcile what he saw as simple-minded barbaric actions with the refined intellectual heights of scientific endeavour. Above all, he could not believe in the principle that science should be applied to developing means to kill people, whatever their political affiliations.

Perhaps fortunately for the war effort, others did not hold the same pacifistic views. Fritz Haber, the director of the Institute for Physical Chemistry and Electrochemistry, the sister institute to the Kaiser Wilhelm Institute for Physics, was an active contributor to the application of science to military purposes. After the importing of raw materials used to make ammonia became almost impossible halfway through the war, Haber's prewar attempts to develop an efficient process for ammonia production were dramatically stepped up. No expense was spared in attempting to synthesise the compound which was so important in the manufacture of fertilisers, animal foodstuffs and, most importantly, explosives. Haber managed greatly to enhance the yield of the process and doubtless played a part in lengthening the war and, consequently, increasing the death toll.

Haber was also employed later in the war to develop suitable gases to be used against the enemy in the trenches. Of course, he was not alone in this task and had his counterparts in the Allied camp.

Of Einstein's contemporaries, Otto Stern, Max Born, Karl Schwarzchild and even Walther Nernst all played their parts in the war effort, either by directly taking part in military action or by applying their scientific talents. Madame Curie was driving Red Cross ambulances during the fighting.

The military authorities never approached Einstein to enlist his help, because shortly after the outbreak of war he had made his position on the matter very clear for all to see.

In October 1914, two months after the war began, what became known as the 'Manifesto of the 93' appeared. This was a document created by the German wartime propaganda machine with the sole purpose of persuading the intellectual community across the world that the action of the government in invading Belgium and broadening the conflict was a justifiable move.

Voices had been raised across Europe by a growing number of intellectuals that the German military action did not tally with the country's great cultural heritage. How could the achievements of Goethe and Schiller have originated from the same people who had discarded international treaties, marched into a neighbouring country and subjected its citizens to unprovoked terror?

The 'Manifesto of the 93' was so named after its creators had secured the names of ninety-three leading German intellectuals from a multidisciplinary background. All the signatories declared their support for the invasion of Belgium and agreed that the move was perfectly acceptable in military terms. Furthermore, they stated the belief that there was no dichotomy between the intellectual and the militaristic actions of the German nation. It was signed by, among others, Einstein's friend and colleague Max Planck.

Within days of its publication, Einstein had signed a counter-manifesto, the 'Manifesto to Europeans', drawn up by a fellow pacifist at the University of Berlin, the biologist Professor Georg Nicolai.[7] This manifesto began:

> Never before has any war so completely disrupted cultural cooperation. It has done so at the very time when progress in technology and communications clearly suggests that we recognise the need for international relations which will necessarily move in the direction of a universal, worldwide civilisation. Perhaps we are all the more keenly and painfully aware of the rupture precisely because so many international bonds existed before.

It ended with a clarion call for unity:

> We ourselves seek to make the first move, to issue the challenge. If you are of one mind with us, if you too are determined to create a widespread movement for European unity, we bid you pledge yourself by signing your name.

The 'Manifesto to Europeans' was a complete failure, managing to secure only two signatories other than Einstein and Nicolai, but it marked a new era for Einstein. In endorsing a document opposing the government in this fashion, he made his first true anti-establishment move, beyond what affected his personal circumstances. It was to create a paradigm for his future reputation.

In the Berlin of 1914, Einstein's action created an official view of him which was to endure throughout the war, as a naive and largely ineffectual anti-establishment figure. The authorities had always known that Einstein was a nonconformist, but they also realised that he had very little real power. Consequently, they left him alone.

Only once in his life did Einstein ever ally himself with a

political party, and this was shortly after the failure of the 'Manifesto to Europeans'. In November 1914, the Bund Neues Vaterland was established by a group of disparate individuals, including the banker Hugo Simon and the businessman Ernst Reuter. The aims of the party were to bring about an early end to the fighting and to set up an international agency which would prevent any future wars. These intentions obviously appealed to Einstein, who quickly enlisted as a member. The Bund Neues Vaterland became increasingly vocal during the war, but was banned in 1916 and continued its operations underground. Only after the Armistice was the party officially allowed to re-form and continue its operations.

Membership of an illegal political party was far from being Einstein's only subversive activity during the war. There is a large body of evidence that he used his limited influence to help attempts to bring about an early end to the war, even if it meant the defeat of his country of residence.

These actions could of course have been viewed by the authorities as treason. But Einstein, who did not consider himself to be in any way German, would not have seen his behaviour in this light. It is a matter of conjecture how much the government knew of his views and his antiwar agitation outside Germany. They may well have been aware of his true purpose in making a seemingly innocent visit to Switzerland in September 1915. The intention of the trip was to call on the famous activist pacifist Romain Rolland, who, Einstein believed, would pass on the scientist's antiwar sentiments to pacifist groups throughout Europe. The government may also have known of the stream of letters Einstein sent to friends abroad in which he constantly called for an end to the war, even if it meant the defeat of Germany.

By his very opposition to the war, Einstein was deriding his adopted government and the policies which the majority felt to be right, although he kept his views very quiet inside Germany. This principled stand would undoubtedly have had an effect on the morale and hopes of the many antiwar campaigners in neutral Europe.

Einstein's antiwar stance was born out of a hatred for militarism, not from any antipathy towards the German people. In

1915, he wrote to his friend Heinrich Zangger, a colleague from his days teaching at the ETH. 'I begin to feel uncomfortable amid the present insane tumult, in conscious detachment from all things which preoccupy the crazy community. Why should one not be able to live contentedly as a member of the service personnel in the lunatic asylum? After all, one respects the lunatics as the ones for whom the building in which one lives exists.'[8]

As the war raged on unabated, Einstein remained separated from his wife and children. Apart from occasional visits to neutral Switzerland, where he found peace and isolation from the craziness of a community at war, he remained in the German capital, where he was putting the finishing touches to the general theory of relativity.

There were growing practical and emotional problems with Einstein's estrangement from his family. As the war progressed, it was becoming increasingly difficult to get money out of the country and into Switzerland. To make things far worse, his personal relations with Mileva were deteriorating.

After a visit to the family during Easter 1916, the relationship between Einstein in Berlin and Mileva and the boys in Switzerland took a sudden nose dive. The couple argued violently and Einstein returned to Germany in a rage, vowing that he would never see Mileva again. Shortly afterwards, twelve-year-old Hans, Einstein's eldest son, stopped writing to his father and Mileva fell ill.

This unhappy situation did not last long. When the animosity over the separation had subsided, the unpleasantness between Einstein and his family lessened. Hans resumed writing to his father and Einstein visited them whenever he found himself in Switzerland.

In the meantime it was decided that Besso, Einstein's closest confidant from college days and a great family friend, should be enlisted as go-between and arbitrator for the couple. Although it was to be over two years before divorce proceedings began, the spring of 1916 marks the point at which the wish for a divorce had begun to take shape in Einstein's mind.

*

The stress of the war, many years of overwork and a catalogue of personal catastrophes were to take their toll on Einstein with unexpected ferocity. In the early part of 1917, he fell into a period of sickness from which he did not fully recover until 1920 and which severely hampered his work.

In many ways, this period of illness could not have come at a worse time, for Berlin in 1917 and 1918 was a war-exhausted shadow of its former self. Before the war, Berlin had been justly proud of its reputation as the capital of the Prussian Republic, a relatively young city which had seen its population increase from an estimated 8,000 in 1688 to around 3 million at the outbreak of the fighting. Years of war had taken the shine off the city. It was difficult for Berliners to find patriotic inspiration in its glorious architecture when their bellies were empty. During the last years of the war, Berlin did not suffer as badly as many European cities involved in the conflict, but Berliners did experience severe shortages of food and raw materials. It was not until 1917 that the war emerged from a stalemate by the Americans entering the conflict.

Einstein was initially diagnosed as suffering from a liver complaint. A short time later he developed severe stomach trouble, which turned out to be an ulcer. Within weeks, jaundice was added to the list of ailments and, accompanying this barrage of physical problems, Einstein was suffering from nervous exhaustion and severe depression.

He had been leading an unhealthy existence. One of his friends who also happened to be a physician, Janos Plesch, described Einstein's lifestyle during the first half of the war thus: 'As his mind knows no limits, so his body follows no set rules . . . he sleeps until he is wakened; he stays awake until he is told to go to bed; he will go hungry until he is given something to eat; and then he eats until he is stopped.'[9]

This description sounds a little exaggerated, but it is clear that during the first half of the war, living as a bachelor in Berlin, Einstein was paying far too little attention to himself and neglecting the fact that he had to keep his body in good health in order to allow his mind to function properly. It is interesting that it was only after completion of the general theory of relativity that

Einstein fell ill, almost as if the let up in the intensity with which he had been tackling his work had allowed illness to take over.

Einstein really needed to be looked after. He needed a supportive wife, or a housekeeper who could regulate his life for him and keep him in working shape, for he was quite hopeless at such things. With Mileva gone from his life and living in another country, things looked bleak. But his life was about to change once again, and this time greatly to his advantage. The woman who was shortly afterwards to become his second wife and lifelong companion appeared on the scene – Elsa Löwenthal.

Elsa's mother was Pauline Koch's sister and her father was a cousin of Albert's father, Hermann Einstein, so they were related on both sides of the family. Elsa and Albert had known each other since childhood as members of the extended family. Elsa had married a merchant while still in her early twenties and had two daughters by him before the marriage had ended in divorce. By 1917, Elsa was living in a roomy and comfortable flat at Haberlandstrasse 5, Berlin, with her daughters Ilse, who was then twenty, and Margot, eighteen.

Elsa was a very homely type; she had a motherly air about her and was, according to family friends, a very warm and affectionate woman with a natural patience and relaxed manner – a very different person from the guarded and somewhat severe Mileva. Elsa had not gained a college education and knew absolutely nothing about science. Perhaps this appealed to Einstein. Elsa proved to be a life-saver for her sick cousin in the dark days of 1917.

Throughout the first half of the year, Elsa nursed Einstein in his rooms when he was unable to leave his bed for months on end and brought him hot meals to help him regain his strength. Owing to the British naval blockade, food supplies were restricted and rationed – there was famine in several Balkan states and starvation as close to home as the outskirts of Vienna. Eating properly in Berlin during 1917 was a problem for most people. For Einstein, ill and unused to being on his own, Elsa's presence may well have made the difference between life and death.

The friendship between Elsa and Einstein developed gradually during the last two years of the war. By the summer of 1917, he had moved to a flat next door to Elsa and her daughters. It was a very practical move and meant that Elsa could spend more time with her patient. Her care and attention gradually worked. By the spring of 1918, Einstein was able to leave the house. A year earlier, when the stomach ulcer had first been diagnosed, he had reportedly lost fifty-six pounds in weight within the space of three months.

With renewed health and Elsa's help, Einstein dragged himself out of the depression which had accompanied the physical symptoms. As his gradual recovery progressed, it became obvious to Einstein and Elsa that they made a very good team. Thoughts of marriage were not far off.

His marriage to Mileva had been in some respects a mystery to outsiders, but the union of Albert and Elsa seemed made in heaven. Certainly Elsa was no match for him intellectually, but Einstein sought companionship in his circle of scientific colleagues and friends in other disciplines. What Einstein needed in a wife was friendship and support. He needed a woman who would look after his domestic needs and ease his path through life while he worked on his science. In return he would give her loyalty, companionship and affection. Elsa could supply all his needs and he felt sufficiently close to her after the traumas of 1917 and 1918 to overcome his distrust of marriage and bitterness at his first attempt.

Of course, in 1918 there was a major hurdle to his future domestic happiness – he was still married to Mileva. Einstein began divorce proceedings.

The logistics of the divorce were complicated and Mileva was highly resistant. She, after all, saw no advantage to being divorced from Einstein and quite possibly resented the relationship which had developed between her husband and his cousin Elsa. However, Einstein was very clear about what he wanted to do and he forced the process through.

Money, of course, was a sticking point and Einstein's success of recent years proved a double-edged sword. It provided the financial means by which a settlement could be reached but at the same time it further fuelled the bad feeling between the Einsteins.

What finally solved the stalemate was Einstein's suggestion that he give Mileva and the children the entire proceeds from the Nobel Prize which he was certain to receive within a few years. This appears to have satisfied Mileva and the settlement was agreed upon, the divorce being formalised in February 1919.

The parting had proved to be less hostile than it might have been after the debacle of 1916. After a few years of coolness, during which the former couple had little contact, Mileva and Albert regained a degree of friendship. Mileva stayed in Zurich for the rest of her life and the family lived comfortably on the interest gained from the Nobel Prize, which was indeed awarded to Einstein in 1922. However, Einstein wrote shortly after Mileva's death in 1948: 'She never reconciled herself to the separation and the divorce, and a disposition developed reminiscent of the classical example of Medea. This darkened the relations to my two boys, to whom I was attached with tenderness. This tragic aspect of my life continued undiminished until my advanced age.'[10]

After the suffering of the past two years, 1919 turned out to be one of the happiest and most important years of Einstein's life. The war had ended on 11 November 1918 and the global carnage had ceased. His general theory of relativity was beginning to gain broader acceptance within the scientific community. In the spring he married Elsa in Berlin and moved into her flat in Haberlandstrasse. He was forty years old.

What Einstein had succeeded in doing, by this midpoint in his life, was to revolutionise the way in which physicists viewed the universe we all live in. He had achieved nothing less than the creation of the first new model of the universe since Isaac Newton, over two centuries earlier. In so doing he had produced an entirely new way of visualising reality, backed up by rigorous mathematical formalism – an achievement sometimes referred to as the greatest intellectual effort of any single human brain.

However, in purely worldly terms, 1919 was merely the beginning. Within months of his marriage, the world would know the name of Albert Einstein. After the disappointment of the prewar efforts to prove his general theory, a new experiment was organised. This time the results would send the scientific world into a flurry which would filter through to the popular press and the general public.

Chapter Eight

The Masterwork

What is it that makes the special theory of relativity 'special'? The term is used as the opposite of 'general', to make it clear that Einstein's first theory of relativity only applies in special circumstances. It applies to bodies that move at constant speeds in straight lines – at constant velocities. A general theory of relativity would be one that explained how the universe works in terms of objects that can also move in curved paths and experience accelerations.

The most important kind of accelerated motion, which makes bodies such as the Earth and the Moon follow curved paths through space, and which describes the fall of an apple to the ground, is caused by gravity. The big gap in the special theory of relativity is that it does not include a description of gravity.

Einstein began worrying about how to extend the special theory to include a description of gravity almost as soon as the special theory had been published, and even before Hermann Minkowski came up with his geometrical description of the special theory in terms of four-dimensional spacetime. Like most of Einstein's great ideas, the general theory of relativity began with a piece of physical intuition, an insight comparable to the picture of a man running alongside an electromagnetic wave moving at the speed of light. In a reminiscence about the birth of relativity that he wrote in 1920, Einstein said that in 1907, while he was working on a review article about the special theory of relativity:

> There occurred to me the '*glücklichste Gedanke meines Lebens*', the happiest thought of my life . . . *for an observer*

> *falling freely from the roof of a house there exists* – at least in his immediate surroundings – *no gravitational field.* Indeed, if the observer drops some bodies then these remain relative to him in a state of rest or uniform motion, independent of their particular chemical or physical nature (in this consideration the air resistance is, of course, ignored). The observer therefore has the right to interpret his state as 'at rest'.[1]

We even know exactly where Einstein was when he had this happy thought. In a lecture that he gave in Japan in 1922, he said:

> I was sitting in a chair in the patent office at Bern when all of a sudden a thought occurred to me: 'If a person falls freely he will not feel his own weight.' I was startled. This simple thought made a deep impression on me. It impelled me toward a theory of gravitation.[2]

The insight led Einstein to a jumping-off point for approaching a general theory, a jumping-off point expressed as the 'equivalence principle'. This says that in order for the acceleration caused by gravity to cancel out the force of gravity (leaving the falling observer in a state of 'free fall'), the acceleration and gravity must be exactly equivalent to one another.

But having arrived at the principle which underpins the general theory in 1907, Einstein found it impossible to make immediate progress with developing a mathematical description of the interactions between bodies that would incorporate the principle. Although he could not know it, the other key ingredient that he would need would be Minkowski's geometrisation of the special theory, which did not appear until 1908 and which Einstein did not immediately welcome even then. At the end of 1907, he put the problem of gravity to one side for more than three years, during which he left the Patent Office and became a fully fledged academic. Much of his time was taken up by giving lectures; and the main thrust of his own scientific endeavours at this time concentrated on developments in quantum theory. There is no evidence at all that he made much of an effort to get to grips with gravity during this time, and plenty of evidence that he did not, as Abraham Pais makes clear in his biography *Subtle is the Lord* . . .

In 1911, however, Einstein returned to the problem of gravity during his sojourn in Prague. For the first time, Einstein realised that the equivalence principle implied an important and measurable effect – that a ray of light passing near to the Sun should be bent. But he still did not have the right mathematical tools to describe what his physical intuition was telling him about the nature of the universe. The next key step came after Einstein returned to Zurich in 1912, at the instigation of his old friend Marcel Grossman.

Einstein had by now fully accepted Minkowski's geometrisation of the special theory. But the geometry used by Minkowski to describe spacetime is itself a special case – it is the four-dimensional equivalent of so-called 'flat' geometry, expounded by Euclid and used by all of us in school, where we learn that, for example, the angles of a triangle add up to 180 degrees and that two parallel lines extend for ever without either crossing one another or moving further apart from one another. Einstein knew that this simple geometrical picture could not be applied to spacetime if all accelerated systems were to be allowed to be equivalent to one another. He needed a more general form of geometry, to go with his more general physics, and (partly owing to his laziness as a student a decade before) he did not know enough about geometry to solve the puzzle. But Grossman did. (He was the one who used to lend Einstein his lecture notes when they were students together in Zurich.) When Einstein turned to him for help, Grossman was able to introduce him to the pioneering nineteenth-century work of the mathematician Bernhard Riemann, who had studied the geometry of curved surfaces, and had developed the mathematical tools to describe this non-Euclidean geometry in as many dimensions as you wanted.

On curved surfaces, the angles of a triangle may not add up to 180 degrees, and 'parallel lines' can cross each other or diverge. On the spherical surface of the Earth, for example, the angles of a triangle add up to more than 180 degrees (how much more depends on the size of the triangle); and two lines of longitude that each cross the equator at right angles (and must therefore be parallel) cross *each other* at the North and South poles.

What Einstein needed, and what Grossman, drawing on Riemann's work, supplied, was a mathematical description of the curvature of four-dimensional spacetime. The appropriate equations for manipulating curved spacetime involve a development of calculus known as tensor theory; just as Newton had needed the tool of calculus itself ('simple' calculus) to develop his theory of gravity, so Einstein, it turned out, would need the tensor calculus to work out *his* theory of gravity. Unlike Newton, however, Einstein did not have to invent the new calculus for himself; it was there, ready made, in the work of nineteenth-century mathematicians such as Riemann, Bruno Christoffel, Gregorio Ricci and Tullio Levi-Cività.

The peripatetic professor had scarcely learned to use the new tool, writing one paper in collaboration with Grossman, before he was off to another new job, the post in Berlin, in 1914. It was there that he put the finishing touches to the general theory, which he presented at three consecutive meetings of the Prussian Academy of Sciences in Berlin in November 1915, and published in 1916. Fortunately, we do not need all the maths in order to get a picture of what the general theory is all about. Leaving out the tensor equations that describe the curvature of four-dimensional spacetime, we can understand exactly what Einstein had achieved using the simple physical pictures that he loved so much himself.

Bending space and time

General relativity is a theory of gravity – the best theory of gravity we have. It was not a response to any observational puzzle (although it did, almost as an afterthought, account for an old mystery about the orbit of Mercury). Einstein was motivated by a deeper philosophical need, the quest for simplicity and unity in nature. If it had not been for Einstein, a comprehensive theory of gravity might not have been developed by physicists for decades, until scientists were pressed to consider the need for such a theory by the discovery of exotic objects with very strong gravitational fields, such as black holes, pulsars and quasars (although it is entirely possible that *mathematicians*, thinking about the meaning of Riemann's work and investigating the nature of curved space-time, might have arrived at something rather like the general theory by a different route)[3].

Astonishingly, when strong-gravity objects were found in the 1960s, Einstein's theory, which had been of real scientific interest only to a few mathematical physicists (unlike his thesis, it had no practical applications), was ready and able to explain their bizarre properties, as a result of abstract theorising by one man half a century before.

Einstein's flash of insight that set him on the path to general relativity, when he realised that a person falling from a roof does not feel the force of gravity, can be developed to explain the key prediction of the general theory, light bending, by imagining that person falling not from a roof but in a lift whose cable has snapped and in which all the safety devices have failed. People in the falling lift will float, completely 'weightless', able to push themselves from wall to wall or floor to ceiling with ease.

We have now seen people in exactly this situation – astronauts in spacecraft, falling freely in orbit around the Earth. In such weightless conditions, objects obey, precisely, Newton's famous laws of motion, proceeding in straight lines unless interfered with by forces. But Einstein had to imagine all the things we have seen for ourselves on television – pencils hanging in midair, liquids that refuse to pour, and so on. Einstein's genius saw all this, and the important point missed by everybody else. If the acceleration of the falling lift, plunging downwards at an ever-increasing speed, can *precisely* cancel out the force of gravity, then that force and that acceleration are exactly equivalent to one another.

The power of this insight – the principle of equivalence – is clear if we imagine the lift replaced by a closed laboratory which is being accelerated through space by a constant force. Everything in the lab falls to the floor, and a physicist who carries out experiments inside will be unable to tell whether the downward force is due to an acceleration or to the force of gravity pulling things down.

Now, said Einstein (in effect), imagine setting up an experiment inside such a lab to measure the behaviour of a beam of light that crosses from one side of the room to the other. In a lab moving at constant velocity, far from any planet or star, the light will travel in a straight line across the lab. But in the accelerated lab, the opposite wall has speeded up and moved forward relative to the light beam in the time it takes the light to cross the room. To the physicist inside the lab it will seem as if the beam of light is bent.

It looks as if there is, after all, a way to distinguish acceleration from gravity. But no, says Einstein. We must keep the principle of equivalence unless it is *proved* false. If the light beam is bent in an accelerated lab (an accelerated frame of reference), then it must also be bent by gravity, and by the exactly equivalent amount.

Since light has no mass, how can it be affected by gravity? The general theory of relativity explained that light bending, and much more besides. The new picture of the universe casts aside the everyday notion of empty space and replaces it by an almost tangible continuum in four dimensions (three of space and one of time); not just a flat (in a four-dimensional sense) and static backdrop for dynamics, as Minkowski used to explain the special theory, but something that can be bent and distorted by the presence of material objects. It is those bends and distortions that provide the 'force' of gravity, bend light beams, and deflect moving objects from straight-line trajectories, a situation summed up by the aphorism 'matter tells space how to bend; space tells matter how to move'.

It is easier to visualise what is going on in terms of a two-dimensional elastic surface. Imagine a rubber sheet stretched tightly across a frame to make a flat surface. That is a 'model' of Einstein's version of empty spacetime. Now imagine dumping a heavy bowling ball in the middle of the sheet. The sheet bends. That is Einstein's model of the way space distorts near a large lump of matter.

When you roll a marble across the original flat sheet, it makes only a tiny indentation, and rolls in a straight line. But when you roll the marble near the bowling ball, the distortion in the rubber sheet caused by the bowling ball makes it follow a curved path. That is Einstein's model for the force of gravity. Objects are simply following a path of least resistance, a geodesic – the equivalent of a straight line through a curved portion of spacetime.

And this explains light bending. The effect is the same for a marble, a planet or a beam of light. When it moves near a large mass – through a gravitational field of force, according to the old picture – it follows a curved trajectory.

The general theory predicted (when Einstein put those tensor equations he learned about from Grossman into the calculation) exactly how much a beam of light should become bent when it passes near the Sun, in order to be exactly equivalent to the bending seen by a physicist in an accelerated frame of reference. The new theory made a clear and testable prediction, that stars observed 'near' the Sun in the sky during a total eclipse (but actually, of course, much further away along the line of sight) would be displaced by a certain amount compared with their observed positions at other times. The prediction was confirmed by observations made during a total eclipse of the Sun in 1919.

During the solar eclipse of 29 May 1919, a team led by the British astronomer Arthur Eddington photographed and measured the positions of several stars which lay in nearly the same direction in the sky as the Sun at the time. Light from the distant stars passed through the region of space affected by the Sun's gravity.

When these positions were compared with the measure positions of the same stars when the Sun was on the opposite side of the sky, using photographs taken six months earlier when those stars appeared in the night sky, Eddington found that they were apparently deflected, each by an amount which depended on the angular separation of the star from the Sun at the time of the eclipse. Light had been 'bent' as it passed by the Sun. These 'deflections' were exactly in line with the predictions made by Einstein's theory.

General relativity has also passed every other test applied to it. One of these concerns what is known as the precession of the orbit of Mercury.

Mercury is the closest planet to the Sun, and orbits where the gravitational field is strong (where spacetime is strongly distorted). Astronomers already knew in 1915 that the orbit has a curious behaviour, which cannot be completely explained by Newton's theory of gravity. As Mercury follows an elliptical orbit around the Sun, the ellipse itself shifts slightly each orbit, tracing out a pattern like a child's drawing of the petals of a daisy.

This shift is exactly explained by general relativity. Anywhere that gravity is weak, general relativity and Newton's famous inverse square law give exactly the same answers to the appropriate calculations. But in a strong field, according to general relativity, gravity deviates from the precise inverse square law. The size of the 'post-Newtonian' effect is just big enough at the distance of Mercury from the Sun to produce the puzzling orbital changes.

The universe at large

General relativity is a geometrical theory. It gives a well-defined physical meaning to a completely specified geometry of matter, space and time. But 'completely' is a key word here. In his search for a unified description of nature, Einstein developed a theory that completely describes the universe – and strictly speaking, *only* describes the complete universe (or a complete universe).

When general relativity is applied to 'local' problems such as calculating the orbits of planets in the Solar System, it is being used in an approximation. In practice, such approximations can be made as accurate as you like, using boundary conditions to join the equations describing a local object like the Sun on to the rest of the universe. But the important point is that Einstein did not have to expand his theory to make it capable of dealing with the whole universe – making it a cosmological theory. General relativity, from its birth, dealt quite happily with the whole universe.

When Einstein tried to describe the simplest possible mathematical model of the universe using his new equations, however, he ran into a problem. At that time, in 1917, the received wisdom was that our Milky Way galaxy was the entire universe, a stable collection of stars. But the equations describing a complete cosmology of space, time and matter refused to produce such a picture. They insisted that the universe must either be expanding or contracting.

The only way Einstein could hold the model universe still, to mimic the appearance of the Milky Way, was to add an extra term to the equations, called the 'cosmological constant'. In 1917, he wrote: 'that term is necessary only for the purpose of making possible a quasi-static distribution of matter, as required by the fact of the small velocities of the stars'.[4] Much later, he referred to the introduction of the cosmological constant as the 'greatest blunder' of his scientific career.[5]

A dozen years later, observers, led by the pioneer Edwin Hubble in California, discovered that the Milky Way was not the entire universe, but simply one galaxy among many millions, and that distant galaxies are all receding from each other. The universe *is* expanding, exactly as the pure equations of general relativity predicted in 1917, when Einstein refused to believe the evidence of his own theory. There is no need for the cosmological constant, and Einstein's equations now provide the basis for the highly successful 'big bang' description of the birth and evolution of the entire universe, which we will discuss in more detail in Chapter Fourteen.

Within the expanding universe, general relativity is required to explain the workings of exotic objects where spacetime is highly distorted by the presence of matter – on the old picture, where large masses produce strong gravitational fields. The most extreme version of this, and one which has caught the popular imagination, is the phenomenon of black holes.

The concept of black holes is so familiar today, as a feature of Einstein's masterwork, that it may come as a surprise to learn that although the name 'black hole' was first used in an astronomical context (by John Wheeler, of Princeton University) only in 1968, the concept goes back more than two centuries, to the work of the British polymath John Michell.

Michell realised that because the speed of light is finite, and because the speed needed to escape from an object (the escape velocity) increases if the size of the object increases but the density stays the same, there must come a point where not even light can escape from the surface of a 'star'. But this would only be achieved by packing a hundred million Suns alongside one another in a huge sphere.

Low-density black holes (with 'only' the density of our Sun, or even less) may indeed exist in our universe, trapping light by their gravitational pull – or, in terms of general relativity, by bending spacetime around themselves so much that it becomes closed, pinched off from the rest of the universe. But there is another way to make a black hole, which was first recognised as a mathematical possibility in the 1930s. If a star keeps the *same* mass but shrinks inward, or stays the same size while accumulating mass, so that density increases, the distortion of spacetime around it increases until, once again, a situation is reached where the object collapses and folds spacetime around itself, disappearing from view from the universe outside. Not even light can escape from its gravitational grip, and it has become a black hole.

The notion of such stellar-mass black holes seemed no more than a mathematical trick, something that surely could not exist in the real universe, until 1968 and the discovery of pulsars.

Pulsars are now known to be the rapidly spinning remains of dead stars, containing about as much matter as our Sun packed into a volume no bigger than that of a large mountain on Earth. Such 'neutron stars' have roughly the density of the nucleus of an atom, and are very close to the critical density at which gravity would overwhelm them and they would collapse into black holes. A neutron star could easily gain enough extra mass to do the trick, by accumulating matter from interstellar space, or by stripping gas from a companion star by tidal forces.

The discovery of neutron stars made the possibility of black holes respectable. In the 1970s, several objects were discovered that might mark the locations of black holes.

An object which emits no light (or anything else) cannot be observed directly. But a black hole orbiting around another star, and swallowing gas that it is tearing off its companion by tidal forces, would be a messy eater. The gas funnelling down into the black hole will get hot, as the particles in the gas are accelerated and bash against one another. Astrophysicists calculated that they would get hot enough to radiate at X-ray frequencies – and X-ray sources in binary systems with the right properties to match those predicted for black holes by the equations of general relativity have now been discovered.

With black holes made respectable by these observations, they were soon invoked to explain another puzzling discovery of the 1960s, the quasars. Quasars are the energetic cores of some galaxies, which produce enormous amounts of energy from a region of space no bigger across than our Solar System.

Allowing matter to fall into a strong gravitational field – converting gravitational potential energy into heat – is the most efficient way to produce energy, apart from the annihilation of particles with their antiparticle counterparts. Dropping a mass, m, into a black hole from infinite distance would release almost half of its rest mass energy, mc^2. If only a few per cent of this available energy is actually released when mass falls into a black hole, the energy needed to power a quasar could be provided by a big black hole which swallows just one or two times the mass of our Sun each year.

The kind of black hole invoked would contain about a hundred million times the mass of our Sun – very much the sort of object envisaged by Michell two centuries ago. And this would be no more than 0.1 per cent of the mass of all the stars in the galaxy surrounding the quasar. Such a black hole could arise simply because too many stars got too close together in the core of a galaxy.

A large concentration of mass will also bend light near it (that is, bend spacetime so that light follows a curved path) even without being a black hole. In some cases, the mass concentration can act as a lens, focusing light from a distant object to produce two (or more) images on the sky. Such gravitational lensing – described mathematically by Einstein in the 1930s – has now been seen in the universe, where multiple images of a single quasar are seen as a result of lensing by an intervening cluster of galaxies.

But the most impressive and complete proof of the accuracy

of Einstein's theory comes from yet another phenomenon, gravitational radiation.

Ultimate proof

The image of matter as solid lumps embedded in a stretched rubber sheet, spacetime, makes the origin of gravitational waves clear. When one of the lumps vibrates, it sends out ripples through the sheet, and these ripples set other lumps of matter vibrating. This is like the way a vibrating charged particle sends out electromagnetic waves which shake other charged particles. But gravitational radiation is very weak; only 10^{-40} times as strong as electromagnetic radiation.

Researchers hope to measure the tiny ripples in spacetime produced by massive objects far from Earth in the near future. But they already have proof that gravitational radiation exists.

A binary system with two very dense stars orbiting rapidly around one another would, according to the equations, be a powerful source of gravitational radiation. A binary system such as this is like an extreme version of a rotating weightlifter's barbell. Viewed in the plane of rotation, this rotation produces gravitational waves in the form of so-called quadrupole radiation, which can be visualised in terms of their effect on a circular ring.

As the wave passes through the ring it is stretched in one direction and squeezed in another at right angles, becoming an ellipse. Then the pattern reverses. The pattern of alternate squeezing and stretching in two directions at right angles is the characteristic signature of quadrupole gravitational radiation.

Three test masses, placed in a right-angle 'L' shape, could detect the passage of such radiation as it squeezes and stretches spacetime. Such systems are now being constructed, with heavy masses placed in evacuated tubes several kilometres long, using laser interferometers to measure their positions to an accuracy of 10^{-18} metres. Researchers expect to detect gravitational radiation with such 'telescopes' during the 1990s.

But even if such gravitational-radiation telescopes do find the ripples in spacetime predicted by the general theory of relativity, this will only confirm what astronomers already know. Just such a system of two neutron stars orbiting one another has been found. It is called the 'binary pulsar'. One of the stars in the system is a pulsar. The other is a neutron star that is not a radio source. They orbit each other every 7.75 hours.

Pulsars are superbly accurate 'clocks', keeping time by the sweep of their radio beams, like those of a lighthouse, as the neutron star rotates. Variations in the pulse rate from the binary pulsar show how the pulsar moves in its orbit – the observed pulse rate speeds up when the pulsar is moving towards us, and slows down when it is moving away. This is a version of the Doppler effect.

The period of the binary's orbit is slowly decreasing. This means that the two neutron stars are getting closer together as time passes. The reason is that the system is losing energy, in the form of gravitational radiation. General relativity predicts that the period of the binary pulsar should decrease by 75 millionths of a second each year; observations are so precise that they show an actual decrease of 76 ± 2 millionths of a second a year.

This is one of the greatest triumphs of Einstein's general theory of relativity, now established beyond any doubt as the best theory we have of gravity and the universe at large.

It was the general theory, and specifically Eddington's announcement of the discovery of the light-bending effect to a joint meeting of the Royal Society and the Royal Astronomical Society in London on 6 November 1919, that catapulted Einstein into the public eye and made headline news.

Part of the appeal of the story lay in the fact that the new theory had been developed by a German-born scientist, based in Berlin, and proved right by a British scientist, based in London, at a time when the two countries were still technically at war. The Armistice had been signed on 11 November 1918, but the Treaty of Versailles which formally ended World War I was not signed until 28 June 1919 – one month, all but a day, after the all-important eclipse. What better example could there be of the way science transcended politics, of reconciliation between enemies, and of hope for the future? Such a description of the response to Einstein's masterwork has heavy overtones of irony today, looking back with hindsight to the events of the following decades. But in 1919 a world weary with war and eager for new sensations read stories in the London *Times* and then other newspapers announcing that space was warped, light did not travel in straight lines, and that the Newtonian idea of gravity had been overturned.

In fact, Newton had not been superseded by Einstein, in the revolutionary sense that those newspaper accounts portrayed. Except under very extreme circumstances, the general relativistic law of gravity is still an inverse square law. The difference is that Einstein's theory explains *why* it is an inverse square law, in terms of the curvature of spacetime. For that matter, light *does* travel in the equivalent of straight lines – geodesics, which are always the shortest distances between two points – through curved spacetime; the point is that it is the spacetime that is bent, not the light. The general theory extends Newtonian ideas about gravity, but it includes both Newtonian gravity and the special theory of relativity within itself as, indeed, special cases.[6]

Probably, though, the headline writers would have ignored those facts, even if they had been aware of them. One of the key elements in the wave of public attention that broke over Einstein at the end of 1919 was the alleged incomprehensibility of his new theory. You may find this surprising, from the sophisticated perspective of the 1990s, especially having just seen how easy it is to get across the basic ideas of the general theory (if not the numbers) using simple geometrical pictures. But the world seemed to want their new hero to be incomprehensibly clever, and one popular story of the time (which Einstein encouraged) was that Einstein himself had once claimed that the general theory was understood by only twelve men (in 1919 it would have been taken for granted that *no* woman could possibly understand the theory).

But perhaps we can understand this public response to the new hero, since something very similar has happened to another cosmologist at the end of the 1980s and in the early 1990s. Stephen Hawking, hailed by many people as the greatest scientist since Einstein, wrote a book, *A Brief History of Time*, that became an international bestseller, in spite of (or, perhaps, *because* of) the widespread story that nobody who lacked a PhD in physics could begin to understand it. Once again, the world had found itself a guru of incomprehensible genius, to admire and set on a pedestal.

Life as such an icon can never be easy, and life for Einstein would never be quite the same after 1919. There had been personal as well as professional turmoil, with his divorce and remarriage during that year, and although, as we shall see in Chapter Ten, his productive scientific work was far from over, the greatest achievement of his professional career was now behind him. Almost imperceptibly, as the world moved into the 1920s, Einstein, now in his forties, began to slip into the role of an elder statesman of science, a father figure to young researchers, and an archetype, in the public eye, of the white-haired, absent-minded professor.

Chapter Nine

Fame but no Fortune

The date was Thursday 6 November 1919; the occasion, a joint meeting of the Royal Society and the Royal Astronomical Society in the main hall of Burlington House, London. An air of excitement and expectancy permeated the Georgian splendour of the room. The scientists who had packed into the hall fell silent as the president of the Royal Society, J. J. Thomson, rose to address the meeting. For a moment he paused and glanced up at the portrait of Isaac Newton hanging high above the gathering. The meeting had been called to make the announcement the scientific world had been waiting for – the findings of Arthur Eddington, recently returned from observation of the solar eclipse at Principe, west Africa. The evidence supported a scientific theory which would alter human perception as dramatically as had Newton's breakthroughs two and a half centuries earlier.

Within twenty-four hours of the announcement, Albert Einstein's theory would become public property, his work described in newspapers around the world and his Berlin home besieged by journalists.

In Burlington House the mood, as described by a member of the audience, the philosopher Alfred Whitehead, was 'that of a Greek drama'.[1] The gathered scientists were fully aware of the historic importance of the occasion. First J. J. Thomson announced the purpose of the meeting and reiterated the importance of relativity in modern physics, declaring that Einstein's theory of relativity was 'the greatest discovery in connection with gravitation since Newton'.[2] Next to take the podium was the Astronomer Royal, Sir Frank Dyson. To a hushed gathering he made the announcement verifying Einstein's theory – that the bending of light by the gravitational field of the Sun observed during the recent solar eclipse did not tally with Newton's theory but coincided almost exactly with Einstein's predicted value.

After the announcements were made, there followed a lively debate about the consequences of the experimental verification. The theory of relativity was still viewed by many as an esoteric and confusing collection of ideas and fully understood by only a few theoretical physicists.

Although relativity had its staunch supporters, it also had its detractors, and those who disagreed with the theory were not simply restricted to the ignorant or those with ulterior motives. During the discussion a number of grievances against the theory were aired, but J. J. Thomson made it clear which side he was on. According to *The Times* of 7 November, Thomson had declared that he was 'confident that the Einstein theory must now be reckoned with, and that our conceptions of the fabric of the universe must be fundamentally altered'.[3] At this point, the eminent astronomer Sir Oliver Lodge walked out of the meeting.

It was Sir Frank Dyson who in 1917, at the height of the fighting in Europe, had noted that the solar eclipse of 1919 would afford a perfect opportunity for the verification of relativity and began to make moves to find the necessary funding. The best view of the eclipse would be from two rather remote parts of the world – the island of Principe, off the west coast of Africa, and, as a backup in case of bad weather, Sobral in northern Brazil. The big problem facing the British astronomers was that Britain was engaged in a bloody and expensive war with Germany. Little money was available for nonmilitary research, and the difficulties in travelling to distant parts of the globe during wartime presented a serious practical hurdle.

Fortunately for Dyson, Eddington and, by association, Einstein too, the war ended in the autumn of 1918, and the funding was found for the expeditions. Eddington said later that the theory might have had to wait for thousands of years to be verified if it had been discovered at some other point in history.

Eddington arrived in west Africa in April 1919 and another group, headed by the astronomer C. R. Davidson, travelled to Brazil. The eclipse occurred during the afternoon of 29 May. Both groups experienced fine weather and the photographic plates were successfully taken.

The photographs were developed abroad and by 3 June, a matter of days after they were taken, Eddington could see that the results looked good for Einstein. However, nothing definite could be announced until the plates had been compared with test prints taken before the expedition set sail. Careful and time-consuming measurements would then have to be made, which meant that it was not until the autumn of 1919 that the results of the experiments could be made public.

The day after the announcement, 7 November 1919, *The Times*, under a headline on page 12 which read 'Revolution in Science – New Theory of the Universe – Newtonian Ideas Overthrown', published a detailed account of the proceedings at Burlington House. Lower down the page there was another article, given the succinct title 'Space Warped', which briefly described the theory.

Thus began the flood of publicity which was to surround Einstein, if later a little less turbulently, for the rest of his life. Not a year passed between 1919 and the scientist's death in 1955 without the name of Albert Einstein appearing in the pages of the *New York Times*. Within twenty-four hours of the London *Times* article, Einstein was selling photographs of himself to journalists, with the money going directly to the starving children of Vienna, still ravaged by the effects of Europe's disastrous war. The photographs appeared in numerous magazines and newspapers around the world and his work was endlessly discussed, dissected, quoted and, more often than not, misquoted by journalists. Albert Einstein had become a household name.

Although Einstein was quick to realise the practical value of his sudden celebrity, he disliked the attention he was attracting as the creator of the theory of relativity. Public awareness of his discoveries was a source of happiness for him, but he was not interested in being thrust into the limelight himself. He soon came to view the fuss being made about his person as rather ridiculous and a distraction from his work.

In a book published in the 1920s, the journalist Alexander Moszkowski describes the response of the public to Einstein's theory of relativity thus:

> Newspapers entered on a chase for contributors who could furnish them with short and long, technical or non-technical, notices about Einstein's theory. In all nooks and corners, social evenings of instruction sprang up, and wandering universities appeared with errant professors that led people out of the three-dimensional misery of daily life into the more hospitable Elysian fields of four-dimensionality. Women lost sight of domestic worries and discussed coordinate systems, the principle of simultaneity, and negatively charged electrons. All contemporary questions had gained a fixed centre from which threads could be spun to each. Relativity had become the sovereign password.[4]

Although this passage sounds exaggerated, other observers agree that it was not too far from the truth. People welcomed the distraction from the war and its after effects. The very themes of relativity transported anyone with the slightest imagination into a world of space travel, distant stars and amazing time-dilation effects not far removed from the fantasies of popular writers such as H. G. Wells and Jules Verne, yet rooted in rigorous mathematics and now shown to be true. Suddenly everyone was a relativity expert. Within the first year of the experimental verification of the theory, over one hundred books had been published on the subject and thousands of articles, critiques, reviews and discussions of Einstein's work had appeared in the newspapers and magazines of the world.

The quality of interpretation of the theory varied enormously. Eddington, an expert on the subject of relativity, lectured to packed houses in many centres of learning and wrote several fine accounts of the basic principles at different levels of complexity. Within a fortnight of the Burlington House announcement, *The Times* published an article commissioned from Einstein himself in which he explained his theory in lay terms. Perhaps not surprisingly, of all those who over the years tried to explain relativity to the general public, it was Einstein who was best able to do this.

Einstein's article was a masterpiece of popular science writing and was lapped up by the general public. However, even with this seemingly innocent article, Einstein managed to create controversy over a humorous addition to the end of the piece in which he commented on his new-found status in the world of popular science.

> The description of me and my circumstances in *The Times* shows an amusing feat of imagination on the part of the writer. By an application of the theory of relativity to the taste of readers, today in Germany I am called a German man of science and in England I am represented as a Swiss Jew. If I come to be regarded as a *bête noire* the description will be reversed, and I shall become a Swiss Jew for the Germans and a German man of science for the English.

The editor of *The Times* was unwilling to cut or change Einstein's personal note but drew a comparison between Einstein's seeming unwillingness to describe himself absolutely and the nature of relativity.[5]

By the early part of 1920, the level of international interest in relativity had reached almost hysterical proportions. A particularly opportunist businessman created the 'Einstein cigar', the creator of the theory of relativity was invited to do a season at the London Palladium, and a leading American magazine, *Scientific American*, offered a $5,000 prize to the writer of the best article explaining relativity. As fate would have it, this prize was won by a senior examiner in the British Patent Office.

However, not everyone was happy with Einstein's theory. Opposition to it in the early 1920s fell into four categories: ignorant ridicule, philosophical incomprehension, resentment by other physicists and political opposition.

First, there were the harmless jibes of the non-scientific media who found the whole idea of relativity alien to what the man in the street would call 'common sense'. Numerous cartoons and spoof articles appeared in the months following the Burlington House announcement, trying to make Einstein out to be some sort of eccentric figure of fun. Relativity and Einstein became by-words for anything that was too complicated for the average person to understand or that seemed slightly flaky. An American journal lumped relativity with:

> the annual sea serpent, the seven-year mutation of our bodies, the jargon of Freud, the messages from Mars . . . Certain troubled spirits, hearing the law of gravitation called in question, do not feel sure that the earth may at any moment slip its Newtonian moorings and go ranging off out of gravitation into the ether – which we now hear does not exist.[6]

The second category comprised a number of philosophers around the world who did not understand the basic precepts of relativity, and endeavoured to create their own interpretations which they then demonstrated to be false. Such critics were merely successful at shooting themselves and each other in the foot.

The third group of critics was equally ineffectual; this was made up of experimental physicists who, to a greater or lesser extent, resented the sudden fame of a theoretical physicist whose ideas they saw as flights of fancy only loosely anchored by experimental verification. These physicists were usually of the type who worked stolidly at their physics in laboratories around the world and saw the subject as rooted in hard facts which could be experimentally demonstrated.

Both these 'experimentalists' and many of the critical philosophers later came round to relativity when they realised its usefulness and better understood its intellectual foundations.

The treatment received by relativity in its early days in the public domain was little different to that handed out to the post-Impressionists when they appeared on the scene a decade earlier. At the first British exhibition of this new movement, the public were, in the words of Virginia Woolf, 'thrown into paroxysms of rage and laughter'.[7] And yet the artists concerned were van Gogh, Gauguin, Picasso, Signac and Cézanne, all later taken under the wing of the establishment.

The only dangerous and completely unreasonable opponents of relativity were those motivated by political reasons.

In the early 1920s many Germans felt that their country had lost World War I by default, by political scheming rather than military inferiority, and a large body of patriotic Germans resented the victorious countries and distrusted anyone who was perceived as having pacifistic or liberal political views. Einstein fitted this bill perfectly. He was becoming increasingly vocal as a pacifist and worker for greater international communication, and his very nationality was open to question. However, most suspicious of all was the fact that he was a Jew.

German anti-Semitism in the aftermath of World War I sprang from the need to find a scapegoat for the country's misfortunes. Although the movement really found its feet with the establishment of the Nazi Party, there was a growing undertow of anti-Jewish feeling in Europe for many years before it was adopted as Nazi policy. The attacks from the far right against the theory of relativity were, perhaps surprisingly, the first time Einstein had experienced raw anti-Semitic feeling directed at him as an individual, and he was both bemused and appalled by it.

The first attack came from a group who euphemistically called themselves the 'Study Group of German Natural Philosophers'. They were a bunch of anti-Semitic fanatics with an eye for personal gain and self-publicity. Their apparent leader was a man called Paul Weyland, of whom very little is known. After a brief burst of fame in the early 1920s, he disappeared back into obscurity.

As an anti-Einsteinian, Weyland had influential friends and through these contacts and the sheer venom of the group's attack on Einstein he presented a very real threat. Weyland's most important ally in his campaign was a Nobel Prize-winning scientist called Philip Lenard. Before World War I, Lenard had been a supporter of Einstein's work on the photoelectric effect, but he opposed Einstein's stance on the war issue and was quick to equate the Jews with the problems facing postwar Germany. There is also little doubt that following the staggering public reception of the theory of relativity, Lenard became consumed with jealousy for Einstein. His reaction was to denounce the theory publicly.

The activities of the Study Group of German Natural Philosophers would have been almost laughable in other circumstances. But in the climate of growing anti-Semitism, they made Einstein's life very uncomfortable for a period. The Group organised a series of meetings all over Germany at which they decried relativity as being a 'Jewish theory' and therefore alien to 'proper' German values and the deeper nature of scientific enquiry. At these meetings, relativity was never attacked on scientific grounds but instead, the creator of the theory was held up as being anti-German and a Jewish revolutionary.

On 27 August 1920, the Group – the 'Antirelativity Company', as Einstein preferred to call them – held a public meeting at the Berlin Philharmonic Hall. In a successful propaganda exercise they managed to pack the theatre, the whole event being a show of false solidarity and single-minded stage management. It was a tiny, amateurish model for the Nazi rallies in Nuremberg some twenty years later. Weyland delivered a speech in which he claimed relativity to be hostile to the German spirit and attempted to engender a feeling among the audience that the woes of the German state were embodied in the person of Albert Einstein, who had created this 'Jewish theory'.

In a brave move, and against Elsa's advice, Einstein attended the meeting, sitting alone in a box he had taken for the evening's entertainment. According to eye-witness accounts, he found the whole event immensely amusing and, as the hyperbole and distortion of his work by the guest speakers became ever more ridiculous, he enjoyed himself more, clapping and laughing out loud at the absurdity of the situation.

Soon after that evening, however, Einstein wrote an aggressive article in the *Berliner Tageblatt*, in which he defended his work and made an attempt to expose the group's anti-Semitism.

The problem with his response was that he really should not have written it at all. By responding, he had, at least in part, dignified the Group's actions and fallen into the trap. If he could have held his tongue and swallowed his pride he would have succeeded in starving them of any sense of satisfaction.

In retrospect, it is very easy to claim that Einstein was wrong to react under provocation. His friends were saddened and disappointed that he had risen to the Group's bait, because they were unworthy of his response, and Einstein later regretted having done so.

Although the 'Antirelativity Company' disappeared into obscurity within a short time of the meeting at the Berlin Philharmonic Hall, their anti-Einsteinian feelings were not isolated. Both inside Germany and in other countries there were individuals and organisations who did not approve of Einstein.

Early in 1920, the authorities of the city of Ulm, Einstein's birthplace, proposed him as a freeman of the city. However, a group within the council vetoed the proposal and the offer was never made. As Einstein's international fame grew, the city made the token gesture two years later of naming a street after him, but it was perhaps more than coincidental that the street lay in one of the tattier areas of the city. When in 1949 Ulm offered to make Einstein an honorary citizen, he turned them down.

Later in 1920, as one of three names proposed for the Royal Astronomical Society's gold medal, he was passed over, despite overwhelming support, owing to the important influence of a small group of anti-Einsteinian astronomers. For the first time in thirty years, no medal was awarded by the society.

Better news came a year later, in 1921, when Einstein was made a Foreign Fellow of the Royal Society in London, just about the highest honour which can be bestowed by the society upon a foreign scientist.

What undoubtedly hurt Einstein more than the withholding of awards and honours was the evident distrust of his colleagues in former enemy countries after the war had ended. It is unclear whether the objections of a minority of British scientists to Einstein was based upon jealousy or political feelings, but they did have influence. It was not until 1925 that Einstein received the Royal Society's highly prestigious Copley medal. The following year he was awarded the Royal Astronomical Society's gold medal denied him six years earlier.

The most damaging effect of this postwar distrust was the refusal of the organisers to allow German scientists to attend the Third and Fourth Solvay Congresses planned for 1921 and 1924. Although Einstein had opposed the German government in World War I, he was cautiously supportive of the more liberal Weimar Republic which had begun in July 1919, and was becoming increasingly angry over the postwar attitude of the Allies. He believed, with many others, that the war reparations demanded by Britain and France were excessively punitive and inhumane. He was also quick to realise the future damage such a harsh approach would create not just in Germany but in the whole of Europe.

The exclusion of German scientists from the Solvay Congress did not extend to Albert Einstein himself. As was pointed out at the time, his nationality was a question of debate and although he had lived out the entire war in the enemy capital, he had experienced and was still experiencing harassment from the authorities. He was a man who had continuously and publicly spoken out against the aggression and had made serious efforts to bring about peace within his limited sphere of influence. Einstein was invited to the 1921 Solvay Congress, but was unable to attend because it overlapped with his first trip to America in March 1921.

By 1924, Einstein deplored the Solvay Congress organisers' continued refusal to allow German scientists to attend, five years after the end of the war. His anger was inflamed by the French decision in early 1923 to invade the Ruhr because an almost bankrupt Weimar Republic had stopped the payment of reparations. Einstein saw this move as cold-hearted and unnecessarily aggressive. He wrote to his close friend

Hendrik Lorentz not to even send him an invitation to the congress.

Happily, by 1927 and the Fifth Solvay Congress, the wounds opened in the scientific community by the effects of World War I had healed sufficiently for the acceptance of German scientists. This was fortunate, because the fifth congress was devoted to the landmark developments of the quantum theory and many of the pioneers of this work were German – Planck, Heisenberg, Born and Einstein. Without their attendance the congress would have been a waste of time.

After the tremendous excitement of the final months of 1919, the beginning of 1920 was a troubled time in Einstein's personal life, for his mother was seriously ill and was moved to her son's house in Berlin to die.

After a life overshadowed by the failed attempts of her husband to create a thriving business, and always living with the fear that they would end up bankrupt, Pauline Einstein had spent her last years in an unsettled pattern. She had stayed with her sister in Hechingen for a time, then kept house for a wealthy banker in Heilbron, followed by a spell with her widowed brother Jakob Koch, from where she was invited to go and live with her daughter Maja and her husband Paul Winteler in Lucerne. It was in Lucerne that Pauline had first fallen seriously ill. Early in 1919 she was diagnosed as having abdominal cancer and was admitted to the nearby Sanatorium Rosenau.

Towards the end of 1919, shortly after the burst of publicity over the experimental proof of Einstein's theories, Pauline decided that she wanted to see her now famous son for the last time and to die in his house. And so she did. In January 1920, she arrived in Berlin accompanied by her daughter and a family doctor and was put up in Einstein's study. She was incoherent or asleep most of the time and kept on pain-killing drugs until her death in March. Afterwards, Einstein wrote to his friend Heinrich Zangger: 'My mother has died . . . We are all completely exhausted . . . One feels in one's bones the significance of blood ties.'[8]

Even for the scientist who had wished from an early age to be detached from the normal flow of life and to isolate himself from all personal contact as best he could, the death of his mother was distressing. He had not maintained a close relationship with his family for the greater part of his adult life and had never been happy about Pauline's attitude towards his first wife, though Pauline had been proved correct about Albert and Mileva's incompatibility. After the couple had separated, Einstein had re-established closer links with his mother, who approved of his second marriage and liked Elsa. Pauline had followed her son's soaring career as best she could, taking pride in his achievements. For his part, unable to spare very much time away from Berlin, especially during the war, Einstein had written to his mother frequently.

Einstein had achieved the great inner strength of a man who saw the world with a detached and distant perspective, yet he formed deep attachments to people and obviously felt their loss when they were no longer there. Although it has been suggested that the development of his theories required a deliberate and nurtured degree of detachment from the comings and goings of human life, he was never able to separate himself completely from the personal and social sphere. He had many friends, but nobody got really close to him. He was often seen to be remote at social gatherings and frequently went off into a vague, dreamy mood in the company of others, when he was quite unaware of what was going on around him.

As Einstein entered his early forties and his fame became greater, he found himself being drawn into areas of life for which he had always exhibited disdain. This became most pronounced in the early 1920s with his involvement with Zionism.

Since his rebellion against organised religion in his childhood, Einstein had shown absolutely no interest in the fact that he was Jewish. To him ethnic identity was meaningless, all national borders and religious institutions were a fabrication. However, from the early 1920s onwards he began to feel a responsibility towards his heritage as a member of the Jewish people. He would continue to remain an atheist, but he realised that he could use his fame and position to improve the lot of Jews around the world.

Einstein had not been particularly aware of his Jewishness until he lived in Prague, where the position of the Jewish community meant that it held the balance of power between the Czech and German communities. Then, when he had moved to Berlin, he was faced with general anti-Semitic feeling as well as the actions of Paul Weyland and his cronies.

If there was one single person who 'converted' Einstein to the Zionist cause, it was Chaim Weizmann, leader of the Zionist movement, who later became a close friend of Einstein. The desire to find a permanent Jewish home was central to the Zionist cause. To this end Weizmann had dedicated himself and realised that, in order to achieve his goal, he needed to have political support from the West as well as large amounts of cash.

Being Russian-born, Weizmann was an ally of the British during the first part of World War I and became director of Admiralty Laboratories soon after the start of hostilities. Trained as a biochemist, he was heavily involved in the development of methods of synthesising chemicals needed in the production of explosives. Through his contact with the Admiralty, he had friends in powerful positions within the government. By 1917, Weizmann had persuaded the British government that there was a need for a Jewish homeland, and by November of that year the Balfour Declaration was issued by the British cabinet. It stated that the British government fully supported the establishment of a national home for the Jewish people in Palestine.

Einstein did not share Weizmann's belief that such a thing was possible in the foreseeable future. He did not like the nationalistic aspect of the Zionists and believed that their world-view was as distorted as that of the people who persecuted the Jews. Einstein was never interested in the power games and intrigues of politics. However, in an approach which was to become a paradigm of his stance on a number of political issues in the years to come, he took the attitude that if a cause had worthy aspects, for the sake of the greater good, faults in the philosophy of that cause could sometimes be overlooked. Thus, although Einstein distrusted the strong nationalist aspect of the Zionist movement and did not believe that the establishment of a Jewish national homeland was at all likely, he did want to use his fame to do some good, and the movement certainly wanted him.

What interested Einstein most of all was the establishment of a Hebrew university in Palestine. For many years he had been witness to the growing difficulty experienced by European Jews in finding places in universities across the Continent. To Einstein this freezing-out of a whole people was one of the more insidious demonstrations of anti-Semitism, for it deprived the Jews of one of the basic human rights – the chance to gain an education. It was not only a problem for the moment. He did not fail to realise that if something was not done about it, it would create damaging repercussions in the future.

It was through the influence of Weizmann and his representative Kurt Blumenfeld that Einstein was finally recruited for the Zionist cause towards the end of 1920.

At the time, Einstein was very unhappy about his position in Berlin. The antagonism of the anti-Semites and the attacks of various groups opposed to his theories and his very Jewishness were weighing heavily on him. During the two years following the verification of relativity, he was as close to leaving the city as he would be for many years. It was largely due to the pleas of his friend and colleague Max Planck, who abhorred the abuse hurled at Einstein, that he stayed. Planck was naturally anxious that Einstein should not leave Berlin University for a foreign centre of learning.

By this stage in his career, Einstein could have taken his pick of lucrative and prestigious appointments at several world-renowned universities. Planck and the German scientific community were fully aware of this and did everything they could to prevent Einstein leaving. Their success was due almost entirely to Einstein's respect for Planck.

Early in 1921, Einstein was persuaded to play his first active part within the Zionist movement. He accepted an invitation to go on a fund-raising tour of the United States with Weizmann, scheduled for that spring.

Einstein had already made a number of foreign visits that year, including trips to Prague and Vienna to deliver lectures on relativity. In Prague he had delivered a public lecture in one of the city's largest concert halls to an audience of 3,000.

A strange incident occurred as Einstein was about to return to Berlin. News that the great scientist was staying at the physics laboratory of the university had spread and a young man clutching a hefty manuscript appeared at the reception, demanding to speak personally to Einstein. After much fussing around and bureaucratic complications, the young man finally had his way. He told Einstein that, based on the famous formulation $E = mc^2$, he was able to prove that a weapon with phenomenal explosive power could be constructed and used for military purposes. According to eye witnesses, Einstein was totally dismissive of the idea and said: 'Calm yourself. You haven't lost anything if I don't discuss your work with you in detail. Its foolishness is evident at first glance. You cannot learn any more from a longer discussion.'[9]

Alas, it was exactly the application of Einstein's equation that resulted in the nuclear explosions over Hiroshima and Nagasaki a quarter of a century later.

The journey to America began in Holland, where the Einsteins boarded the *Rotterdam*, joined by the Weizmanns. They set sail on 21 March 1921.

If Einstein had thought that the reception his work had received upon verification in 1919 had been excessive, the scenes awaiting him in New York were beyond his wildest dreams. He never could understand why the general public were making such a fuss about him as a person, or how the masses, usually bereft of any interest in science, could overnight develop a voracious hunger for such a new and exotic theory as relativity.

Reporters and the interested public besieged the Einstein-Weizmann party as soon as the *Rotterdam* reached New York harbour. Photographers came aboard the ship before the passengers had managed to disembark and Einstein was met by a barrage of questions about his work by popular journalists who wanted packaged, easy-to-understand and instant descriptions of what exactly was meant by relativity. After the hubbub of the journalists and celebrity-seekers, the party were officially met by

Mayor Hylan and Fiorello La Guardia, president of New York City Council.

Einstein was invited on the tour for three reasons. First, he was there as a celebrity guest and crowd-puller. Weizmann realised that Einstein's presence would attract great media and public attention as well as the large number of wealthy and powerful American Jews he so needed for his mission. The second reason, and for Einstein himself the primary one, was to help raise funds for the Hebrew university. A third reason was the opportunity the tour offered for him to lecture on relativity, thus enabling him to present his work to a larger public.

The visitors were fêted by the authorities and attended one banquet after another, arranged in advance by the highly efficient organisation headed by Weizmann. Einstein spoke little on the subject of Zionism. This was partly due to his own lack of conviction – he felt unable to represent the cause honestly – and partly because he was not a very good speaker on matters outside scientific areas. Weizmann was also wary of the scientist making any public statements that could cause the Zionist movement more harm than good.

Einstein did, however, speak passionately and effectively on the subject of the Hebrew university and was successful in reaching the pockets of his audience. They could not really fail to be moved by the sheer presence and obvious enthusiasm of this little man who gave the impression of having been dragged out of his dark and musty study and thrust into the full glare of America. His relativity lectures were also well received. Because his English was so poor, Einstein employed the services of a translator, the mathematician Morris Raphael Cohen. This was undoubtedly a good move considering the complexity of the material. Relativity was perceived as almost impossible to understand without the added problem of having it described in a heavy foreign accent.

After New York, the party visited Princeton University in New Jersey, where Einstein was awarded an honorary degree and delivered a series of lectures. From there he travelled with Elsa to Chicago, where he met the eminent American physicist Robert Millikan, who a few years earlier had been responsible for the experimental verification of Einstein's theoretical work on the photoelectric effect. The visitors then moved on to Cleveland, where they were met by a 200-strong motorcade, and the majority of the Jewish shops were closed for the occasion.

Shortly before leaving Germany, Einstein and his wife had been invited to visit Britain on their way back to Europe. The invitation had come from Sir Henry Miers, vice chancellor of Manchester University, asking Einstein to deliver the Adamson lecture, which had in previous years been given by such luminaries as the Nobel laureate J. J. Thomson, a former president of the Royal Society. Einstein had gladly accepted the invitation and others soon followed from eminent institutions, including one from King's College, London. He also received the personal invitation of one of his great admirers, Viscount Haldane, who offered to act as his host during the scientist's stay in London.

The Einsteins arrived in Liverpool on 8 June 1921 aboard the White Star liner *Celtic*. The couple were exhausted from their journey and the exertions of their lengthy tour of the States. Fortunately the reception Einstein received in Britain, although enthusiastic, was of a far more respectful, subdued nature and more akin to what he had expected. However, after his experiences in America, where he had grown increasingly disdainful of the attitude of the popular press, he now refused to discuss relativity with journalists or interviewers, even if they were more polite than their American counterparts.

The Adamson lecture, delivered in the main hall of Manchester University, was an unqualified success and before leaving the city for London he was made a doctor of science, the first such honour to be awarded to a German since the outbreak of World War I. Einstein also took the opportunity to give a talk on the subject of Zionism at the university, reiterating his keen interest in establishing a Hebrew university for Jewish scholars.

Upon their arrival in London, the Einsteins were met by their host, Viscount Haldane of Cloan, who had been secretary of state for war at the outbreak of hostilities in 1914 and was a former Lord Chancellor. Haldane had close links with the German nation and was a great admirer of the country's cultural heritage and scientific prowess; in particular, although not a scientist himself, he was a huge fan of Einstein and his work.

Haldane was primarily an amateur philosopher, a man of great culture and broad education, intelligent, knowledgeable and possessed of a healthy respect for pure science. In the study of his family home in Cloan, he is said to have had only two pictures hanging on the walls, one of his mother, the other of Einstein.

Among the deluge of books on the subject of relativity which swamped the bookshops immediately after the verification of the theory, one of the best was a work by Haldane. Its subject was really the relationship between knowledge and reality, and it gained the enthusiastic interest of such scientists as Sir Oliver Lodge and Sir Arthur Eddington, who had very differing views on the subject of relativity.

Haldane was perfectly suited to guide his foreign guests through the web of English society at a time when the Establishment were still unsure how to respond to a scientist from a former enemy state. Haldane was wealthy and had many friends from the academic, political and media spheres.

After a meeting with Eddington at the Royal Astronomical Society at Burlington House, their host escorted Albert and Elsa to his London home at Queen Anne's Gate. When the Einsteins arrived, Haldane's daughter fainted from the excitement of meeting the great German scientist.

For the couple's first evening in the capital, having fended off suggestions of a glitzy party thrown by Lady Cunard in Einstein's honour, Haldane had organised a dinner party at his house. The guest list was both illustrious and eclectic. It included figures from the arts, politics, the military and science as well as representatives from the religious establishment. Those present included the scientists Eddington and Whitehead; the playwright George Bernard Shaw; the dean of St Paul's; General Sir Ian Hamilton, veteran of Gallipoli; Harold Laski of the London School of Economics; and the archbishop of Canterbury, Randall Davidson, who was apparently given a crash course on relativity by J. J. Thompson before the dinner but was still none the wiser.

By this time Einstein had grown adept at sidestepping enquiries about relativity from interested nonscientists. According to various accounts of the evening, both he and Elsa spent the major part of the dinner doing just that. It was an unavoidable consequence of Einstein's fame through a subject so esoteric that he should find himself plagued by requests for explanations. He very quickly realised that if he was to escape becoming entrenched in a discussion which could not be fully understood by the other party, he had to do his best to avoid it in the first place.

Thus, when the archbishop of Canterbury tried to engage Einstein in a discussion on the subject of his theories, he found him less than forthcoming. Davidson had asked whether Einstein thought it true that relativity would create a great difference to our morale. Einstein wearily replied: 'Do not believe a word of it. It makes no difference. It is purely abstract – science.'[10]

On such occasions the uninitiated often tried to find a way through what they saw as a quagmire by aligning relativity with mysticism. Einstein never saw himself as a mystic, and despite the attempts of latter-day science writers to make comparisons between his work and Eastern religions and philosophies, the creator of relativity would have found such links laughable.

It is often the fate of physicists working on the edge of science, particularly cosmologists and particle physicists, to be branded as mystics and to have their theories discussed in the same breath as the Bhagavad-Gita. In recent times, many cosmologists have had to endure the same misinterpretations and misunderstandings of what they are trying to achieve in physics. Many find such links offensive; Einstein was more magnanimous about it, perhaps because such cerebral leaps were a new thing in the 1920s.

Aside from a talk at King's College in the Strand, the rest of the Einsteins' visit to England was spent relaxing and meeting a string of important and famous people. Haldane introduced them to the Rothschilds, Prime Minister Lloyd George and Lord Rayleigh and they were taken on a visit to Oxford by Einstein's friend and fellow scientist Frederick Lindemann, Viscount Cherwell.

Before the lecture at King's, Viscount Haldane had gone with Einstein to Westminster Abbey, where they had laid a wreath on Isaac Newton's grave. Introducing the lecture that evening, Haldane mentioned this and said: 'What Newton was to the eighteenth century, Einstein is for the twentieth.'[11]

Einstein's visit to Britain was widely seen as a great success, not least because his presence could begin to heal the rift the war had created between the UK and Germany. As an indication of Einstein's fame and popularity, *The Times* devoted ten articles to him during his two-week stay.

For Einstein the year 1922 was to bring mixed fortunes. It brought great recognition within the scientific establishment, varied receptions in foreign lands and worrying developments in his home city of Berlin.

Early in the year, Einstein was invited to deliver a lecture at the most eminent scientific academy in France, the Collège de France in Paris. Although the talk was on purely scientific matters and political subjects were not mentioned, it proved to be the first occasion on which Einstein experienced any form of anti-German feeling directed against him during a foreign visit.

The invitation for the talk came from the physicist Paul Langevin, who was to become a lifelong friend of Einstein's. Langevin was a great supporter of relativity and was active in publicising an accurate representation of the theory. In many ways he was as much of a political idealist as Einstein and shared the belief that political and social change could be brought about by the application of logic.

France had suffered greatly during the war and the resentment felt by many sections of the French population towards their neighbours had diminished very little in the four years since the ending of hostilities. This anger and resentment was also reflected within Germany. Many Germans hated the French as much as the French hated them. Consequently, by agreeing to deliver a lecture in France, Einstein had laid himself open to attack from both sides.

Realising that Einstein might be facing real physical danger during his visit, Langevin had decided to meet his colleague at the Belgian border and escort him personally to Paris, all the while keeping in contact with the French police, who were on the lookout for troublemakers. There had been threats from political extremists, including warnings from a movement calling themselves the 'Patriotic Youth', who had threatened to interfere with Einstein's visit and to disrupt his lecture.

As the train carrying Einstein and Langevin approached the Gare du Nord in Paris, Langevin received news that a large group of noisy youths had gathered at the station and that it would be best if Einstein could be whisked away through a side entrance to the station. The train was stopped short of the station, the famous guest disembarked and was quietly escorted out of the station and into the metro before anyone at the station had become aware of what was happening. Only some time later was it discovered that the gathering at the Gare du Nord had in fact been a group of young Einstein supporters from the university who had come to welcome him to the city. The leader of this group was none other than Paul Langevin's son.

Thanks to the wise precaution of making entry to Einstein's lecture at the College de France by ticket only, and ensuring that only the genuinely interested should be sold tickets, the lecture went off without a hitch. Einstein delivered the lecture in French, his second language, which he spoke rather haltingly and carefully. Einstein's lecture was well received by the gathered academics, but the politically active used the occasion of the visit to continue grinding their particular axes and to stir up international distrust.

The attacks came from both sides. Einstein was not invited to speak before the French Academy of Sciences after thirty members stated that if he was invited to give a talk, they would walk out in protest. Then, after Einstein had returned to Berlin, he was dismayed to find, at his next attendance at a session of the Prussian Academy of Sciences, that many of the seats either side of him remained empty.

The French and German media were often caught in the difficult position of not knowing on which side of the fence to fall. They too were fully aware of the vitriol of certain factions of the public of both nations, but Einstein was a much admired figure in the world of science and a popular character with the majority of the public. One French newspaper asked whether, if a German had discovered a cure for cancer instead of relativity, members of the Academy would have to wait for Germany to join the League of Nations before the remedy could be practically applied.

By the summer of 1922, fears concerning the actions of political extremists increased dramatically when the German foreign minister Walter Rathenau was assassinated by right-wing extremists in Berlin. At the trial of the assassins it became clear that Rathenau had been murdered not so much for his political stance as for the fact that he was a Jew. The murderers apparently did not know what he did, but were only aware of his ethnic heritage.

Rathenau's assassination was not an isolated incident. Earlier in the same month, Herr Scheidemann, a former German prime minister, had narrowly escaped a similar attack, and within days of the Rathenau killing, the prominent publicist Maximilian Harden was seriously wounded in a foiled assassination attempt. Both Scheidemann and Harden were Jews.

The inevitable question was whether Einstein might be the next target of the assassins.

The anti-Semites and the fanatical nationalists both saw Einstein as the embodiment of all that was wrong with Germany. Einstein had already experienced bad feeling little short of actual violence on a number of occasions since he had achieved international fame. During his visit to the United States in 1921, an extremist had been charged and fined when he had publicly offered a reward for the murder of Albert Einstein. Evidently there was no shortage of individuals and organisations who wanted to see the great physicist dead.

The murder of Rathenau undoubtedly shocked Einstein. The foreign minister had been a personal friend, and indeed Einstein had advised him not to take the appointment because of the political atmosphere of the time. Einstein was not the only one to see the beginnings of a national catastrophe in these events. Attacks on Jews would become commonplace within a decade and would finally lead to the tyranny of the Nazis and the horrors of the concentration camps.

If Einstein was shocked, Elsa was terrified for her husband's safety. Later that summer Einstein was due to visit the Netherlands to deliver a series of lectures on relativity. Elsa was convinced that assassins were going to attempt to kill him before he left the country. Without ever letting on to Einstein, she persuaded the chief of police to protect him by employing dozens of plain-clothes policemen to follow Einstein around and to escort him on his foreign trip. Einstein had no idea that many of the other passengers in his train carriage en route to Holland were his undercover protectors.

After Einstein's return from Holland, the atmosphere in Berlin was as oppressive as ever. The day of Rathenau's burial had been declared an official day of mourning and all academic and cultural institutions were closed. However, Einstein's most bitter opponent, Philip Lenard, made a very public show of continuing with his lectures as normal.

As the summer drifted into early autumn, there came temporary relief from the tensions of life in Germany. In October Albert and Elsa embarked the SS *Kitano Maru* in Marseille bound for Japan and the start of an extended tour of the Far East.

The Einsteins arrived in Japan on 20 November 1922. He was scheduled to deliver a series of lectures, to both public and academic audiences. The Japanese received Einstein with enormous enthusiasm, blending respect for the creator of the theory of relativity and genuine interest in the subject with a relaxed attitude to publicity without what Einstein saw as the silly hyperbole of the Western media; a stance which altogether delighted the Einsteins.

One amusing anecdote of their trip to Japan grew out of Einstein's fears for the comfort of his audience. After his first lecture in the country, Einstein realised that the whole talk, including translation, had lasted over four hours. He was staggered by the fact that the audience had remained quiet and attentive throughout. En route to the next venue the following day, he managed to cut the lecture to a more comfortable two and a half hours.

After the second lecture, he noticed that his hosts seemed

disturbed and preoccupied by a conversation among themselves conducted in Japanese, an unusually impolite thing to do. When Einstein finally asked them what the problem might been, he was told that the organisers of the second lecture had been offended by the fact that he had shortened the lecture from the original given at his first stop.

It was during his visit to Japan, at the end of November 1922, that Einstein learned that he had been awarded the 1921 Nobel Prize for Physics. Although the award of the prize was considered by most of his fellow scientists as being long overdue, he was nonetheless delighted.

He had been proposed for the prize as far back as 1910, but the Royal Swedish Academy of Sciences, who make the award, were faced with a number of problems concerning Einstein's scientific work. In his will, Alfred Nobel stipulated that the award must be given to a scientist who has made a discovery from which humankind has derived a great use. The problem facing the Nobel committee in Stockholm was the debatable practical merits of relativity as it could be viewed at that time.

In order to circumvent this difficulty, with the persuasion of a large body of European physicists, it was decided that Einstein should be given the prize for his work on the photoelectric effect, which had already found numerous applications and was undoubtedly a benefit to humanity. In the official citation, the award is said vaguely to be 'for the photoelectric law and his work in the field of theoretical physics'.

Predictably, Einstein's enemies leaped on this as an opportunity to discredit him, and the vitriolic Lenard wrote to the Swedish Academy declaring his belief that the prize had been given merely to attempt to restore Einstein's lost prestige without compromising the Academy.

However, the Royal Swedish Academy of Sciences did not see it that way; neither did the vast majority of scientists, including those who did not fully support relativity but felt that the award was long overdue and thoroughly deserved for Einstein's other substantial contributions to the body of scientific knowledge.

Halfway through his tour of the Far East, Einstein was not available to receive the award, but he was represented at the festivities in Stockholm by the German envoy Rudolf Nadolny. Even in this detail there lay controversy. For the official citation the scientist was declared to be a German. This surprised the Swiss embassy because Einstein was, at that very moment, travelling on a Swiss passport. Einstein also objected to being described as a German, an assumption based on the German citizenship supposedly bestowed upon him by his election to the Prussian Academy of Sciences.

A compromise was reached whereby Herr Nadolny accepted the prize for Einstein in Stockholm and the Swiss ambassador to Germany delivered the award to Einstein in Berlin upon his return from the Far East.

This latest controversy over Einstein's nationality prompted the authorities to sort out the matter. After a number of documents and letters had passed between Einstein and a procession of increasingly senior civil servants, it was finally established that Einstein had indeed become a German citizen when he was elected to the Prussian Academy of Sciences. However, it was also made clear that he was entitled to claim Swiss nationality and use a Swiss passport if he so wished. This is how the situation remained until, in 1933, Einstein handed in his German passport and left German soil for the last time.

As promised at the time of their divorce, Einstein gave the cash element of the Nobel Prize to his first wife, Mileva. In 1923 it amounted to 121,572 kronor (then about $32,000), a sum which proved to be quite adequate for the former Mrs Einstein to live in material comfort for the rest of her life.

As well as being the year in which Albert Einstein finally received the official stamp of merit from the scientific establishment in the form of the Nobel Prize for Physics, 1922 in particular and the decade following World War I in general will be remembered as an important era in a number of spheres of human activity. It was a time of great reawakening in the arts, literature and music, a period when many traditional approaches to human thought and expression were being superseded by new and innovative work. James Joyce's *Ulysses* and T. S. Eliot's *The Waste Land* both appeared in 1922; revolutionary works of art as different from the literary tradition from which they had sprung as relativity had been within the realms of science. The early 1920s were marked by the growing recognition of the 'new music' of Arnold Schöenberg and the 'new art' of such painters as Pablo Picasso, the Surrealists and the German Expressionists; the continuing emergence of 'new philosophies', as proposed by such giants as Ludwig Wittgenstein; and the development of new and staggeringly original scientific theories.

Upon Einstein's return to Europe, he was not only to experience a renewed and eventually intolerable development of anti-Semitic feeling, but he was also to find himself at the centre of the second of the two great scientific developments of the twentieth century. It lay just around the corner as he and Elsa sailed back to Europe early in 1923.

Chapter Ten

Quantum Pioneer

The award of Einstein's Nobel Prize was for his pioneering work in what is now called quantum physics, not for his two theories of relativity. So far, we have largely ignored Einstein's quantum contributions in the years following the *annus mirabilis*, while we explained the two theories for which he is now best known. But quite apart from the interpretation of the photoelectric effect in terms of light quanta, the contributions Einstein made to the development of quantum theory in the first quarter of the twentieth century rank him head and shoulders above all but a handful of other quantum pioneers.

Between 1907 (when he stopped, temporarily, trying to generalise the theory of relativity) and 1911 (when he returned to the puzzle of developing a theory of gravity), some of Einstein's most important work was on the nature of specific heat. This lacks the glamour and wide implications of some of his more famous work. But it was, in fact, through Einstein's explanation of details of the way in which bodies absorb heat, in terms of the kind of quantised 'oscillators' used by Planck in his theory of black-body radiation, that quantum ideas were first introduced to a wide audience of physicists, who then began to take the notion seriously.

Before this work, the quantum idea had only really been used in one context, the rather esoteric question of the nature of black-body radiation and the resolution of the ultraviolet catastrophe. But like his work on the behaviour of particles suspended in a fluid, Einstein's work on specific heats has practical value; just as the thesis work encouraged physicists to take the idea of molecules seriously, the specific-heat work encouraged physicists to take the quantum idea seriously.

In the same years, up to 1911, Einstein worried a lot about the nature of light, and the physical meaning of the interpretation of the photoelectric effect in terms of light quanta. He was careful never to say that light *is* a stream of particles, but only that it behaves *as if* it is a stream of particles. He was well aware of the paradox that while the photoelectric effect seemed to show light behaving as a stream of particles, the familiar Young's slit experiment still showed light behaving as a wave. What is more, although Planck's quanta are necessary to explain the nature of the black-body spectrum for high frequencies (short wavelengths), it turns out that they are not necessary at all to explain the low-frequency (long wavelength) end of the black-body spectrum. At long wavelengths, even black-body radiation seems to behave perfectly in accordance with the classical description of an electromagnetic wave. Black-body radiation itself seems to contain a mixture of particle and wave properties.

One thing was clear in all this. An understanding of the nature of light was inextricably bound up with developing an understanding of the nature of atoms, and the way in which they could absorb and emit light. It is ironic that it was just at the time that Einstein (who had helped to establish the reality of atoms and was responsible for the concept of light quanta) was deeply embroiled in developing his general theory of relativity, from 1911 onwards, that the Dane Niels Bohr was developing the first reasonably accurate physical description of the atom, using quantum ideas. But once Einstein had got the general theory off his chest, he soon returned to the subject of light and atoms, with a burst of productive activity that, though it may not even rank in the top three or four of his scientific achievements, pointed the way towards the development of lasers (forty years ahead of their time) and ushered physicists to the brink of a fully developed theory of quantum mechanics based on the fact that *all* subatomic entities, not just photons, can be regarded as having both particle and wave characteristics – so-called 'wave-particle duality'.

Bohr's atom

By 1911, the basics of the modern understanding of the structure of atoms had been worked out by the New Zealand–born physicist Ernest Rutherford, who was by then working in England, at Manchester University. Experiments had shown that when tiny subatomic particles[1] were fired at a very thin sheet of metal foil, most of the particles went straight through the foil, but a few were bounced back in the direction they had come from. Rutherford explained this by saying that most of the matter in an atom is concentrated in a tiny central nucleus, which has a positive electric charge. This is surrounded by a relatively huge volume (the proportions are roughly the same as for a grain of sand in the middle of the Royal Albert Hall) in which all the electrons associated with the atom swim. Electrons (discovered only in the 1890s, remember) are very light particles, compared with nuclei, and each has one unit of negative charge. Together, the charge of all the electrons balances the positive charge on the nucleus and makes the atom electrically neutral. A positively charged alpha particle would brush through the electron cloud like an artillery shell through a fog bank; but if an alpha particle happened to be heading directly towards a positively charged nucleus, then, since like charges repel, it would be bounced back from whence it came.

This poses an immediate puzzle. Just as like charges repel, opposite charges attract. An atom consists of a positively charged nucleus, surrounded by a cloud of negatively charged electrons. So why don't the electrons fall into the nucleus?

If they did so, they would radiate energy continuously as they fell. All material objects – everything made of atoms – would disappear, literally in a flash. Bohr, who came to work in Manchester in 1912, resolved the puzzle by suggesting that the electrons can occupy only certain well-defined orbits around the nucleus. They could not spiral in towards the nucleus, radiating energy continuously as they did so, because they were only allowed to emit whole pieces of energy – whole quanta – not the continuous radiation required by classical theory. An electron can jump from one orbit – one 'energy level' – to another, emitting or absorbing the appropriate quantum of energy as it does so. But it cannot jump to any in-between state, because there are no in-between states. And, said Bohr, there was a limit to the number of electrons 'allowed' to occupy any one orbit. Electrons are kept apart from one another, with a limited number allowed on each energy level, and cannot fall into the nucleus.

All this could be explained even if the energy itself – light or other electromagnetic radiation – came with any amount of energy, rather than being quantised. But, fresh from his triumphant completion of the general theory, in 1916 Einstein found that he could explain Planck's formula for the black-body spectrum in terms of the interaction of real photons, particles of light, with atoms constructed in the way Bohr had described.

Obviously, in a real atom there will be many possible energy levels, and many different ways in which electrons can jump about between them, absorbing or emitting different, but precisely determined, amounts of energy. This explains the nature of the spectrum of a hot object or gas, which is crossed by characteristic dark lines of absorption or bright lines of emission. It was, indeed, from the measurements of such spectra that Bohr arrived at an equation describing the energy levels of electrons in atoms. But in order to explain Einstein's breakthrough in 1916,[2] it is simpler to imagine an atom which possesses just one electron, which jumps between only two energy levels that are separated by an energy gap which exactly matches the energy of the photons of one particular frequency (one particular colour) of light. Einstein considered the behaviour of such a system (and much more realistically complicated atoms) in thermodynamic equilibrium, but with photons and atoms interacting with one another in accordance with the quantum equivalents of the rules of statistical mechanics. Once again, there is a strong link with Einstein's earliest research endeavours, back in the early years of the century, even though his conclusions would be revolutionary.

There are three possibilities. An atom in which the electron is in the lower energy level may absorb a photon that comes its way, so that the electron moves up to the higher energy level. An atom that has an electron in the higher energy level already may spontaneously emit a photon, at random (this photon may then, of course, be absorbed by another, 'empty' atom). And if a photon interacts with an atom where the electron is already in the higher energy level, then the electron will immediately fall down to the lower level, emitting another photon which proceeds alongside the first. This is called stimulated emission of radiation.

In this work, Einstein explained the nature of the 'jumps' of electrons between orbits in the Bohr model of the atom, and he derived Planck's formula for the black-body spectrum in a new way. Now, he accepted the idea of light as being made up of particles – photons – and in a related piece of work at that same time he introduced the idea of photons carrying momentum. For a photon with a particular frequency f, remember, its energy E is given by hf, and, Einstein now proved, it carries a momentum given by dividing its energy by its speed – E/c. This strengthening of the concept of particles of light moved physics closer towards the notion of wave-particle duality. The reality of photon momentum was established experimentally by the American physicist Arthur Compton in 1923, when he measured the transfer of momentum from photons to electrons in collisions which 'scatter' the electrons in different directions – a process now known as the Compton effect, and regarded at the time (and since) as one of the most important proofs of the reality of photons and the accuracy of quantum theory as a description of the subatomic world.

But Einstein had also, in passing, laid the foundations of a new phenomenon, one that would come to practical fruition only in the 1950s, and which now forms the basis of a multimillion-dollar industry. The notion of stimulated emission of radiation led to the idea of the laser, which gets its name from the acronym for the expression Light Amplification by Stimulated Emission of Radiation. The 'stimulated emission' bit is directly based on Einstein's 1917 paper; the 'light amplification' comes in because in a laser a weak source of light is amplified as the photons move past atoms that are primed in their high-energy state, triggering the emission of a cascade of photons, all with the same energy and frequency, to produce an intense beam of light of a single pure colour. The realisation that each absorption or emission of a photon (each jump of a single electron from one energy level to another) involves a pure frequency, and therefore a pure colour, was also Einstein's.

Like his thesis, Einstein's 'quantum theory of radiation' paper is one that has been extensively cited in recent decades, as lasers have come into their own.[3] The uses of lasers are now legion, but it is perhaps worth reminding you that every compact-disc player uses a laser beam to scan the CD itself. The science behind the CD player stems directly from one of Einstein's lesser-known insights – something he came up with within months of the completion of the general theory of relativity, when he might have been expected to be resting on his laurels for a time.

But in the midst of his delight at the success of his marriage of the quantum theory of light with the quantum theory of atoms, there was one thing that troubled Einstein, and that would nag away at him until, eventually, he would reject the entire quantum theory that he had done so much to found. This was the fact that an atom seemed to emit a photon ***at random***. How did an atom 'know' when it ought to emit the photon? On what basis did it 'decide' to do so? And as well as the problem of when the atom will emit a photon, there is the similar problem of which direction the photon will be emitted in. The property of momentum, which Einstein had now shown to be possessed by all photons, carries with it an indication of direction. Once again, the equations said that the direction in which an atom emits a photon is chosen entirely at random during the process of spontaneous emission.

In the very paper announcing his exciting new results concerning the quantum theory of radiation, Einstein wrote that it was 'a weakness of the theory . . . that it leaves time and direction of elementary processes to chance'. At the time, he clearly expected that improvements to the quantum theory would remove this element of chance, so the difficulty was no more than a small cloud on the horizon. Eventually, that cloud would grow, in

Einstein's view, to overwhelming proportions. But not before he had made two more major contributions to quantum theory.

How to count photons

After the news about the general theory of relativity hit the headlines in 1919, Einstein became busier than ever. He was in demand as a lecturer and received invitations to visit academic institutions around the world, travelling widely in Europe and also visiting, as we have mentioned, the United States, Palestine and Japan. This, remember, was in the days before global air travel, and the longer journeys were made by ship. In addition, as we have seen in Chapter Nine, the turmoil in Germany following World War I gave Einstein, along with everybody else, plenty to worry about apart from scientific work. From 1919 to 1924 he was not particularly productive, by his own previous standards, although he did publish, in 1922 and 1923, his first papers on the problem of a unified field theory. This was to be an endeavour – ultimately fruitless – that would occupy him for the next three decades, until his death in 1955.

In the early 1920s, physicists knew of only two forces (two 'fields') which controlled the behaviour of particles and waves in the universe. Gravity, which operates on the large scale and holds the universe itself together, and electromagnetism, which deals with the interactions between charged particles and also describes the behaviour of light. Nothing could be more natural than that Einstein, who had done so much to improve the understanding of both gravity and light, should attempt to find a scientific theory – a set of mathematical equations – that would describe both these phenomena in one package. In spite of the intensity of his efforts and the length of time for which Einstein beavered away at the problem, however, his attack on it proved to be taking him up a blind alley, and we shall not discuss it in detail here, although the modern version of unified field theory is at the heart of physicists' endeavours to explain the workings of the universe today, and will be mentioned again in Chapter Fourteen.

Einstein's growing status as the most famous physicist in the world also meant that he received a huge number of letters, both from other scientists and from the general public. Dealing with all these communications was a time-consuming process. But in 1924 one of those letters, from a young Indian scientist called Satyendra Bose, brought with it the key to the next step forward in the development of quantum theory.

Bose, who was working at Dacca University, had found yet another way of deriving Planck's equation for the spectrum of black-body radiation. But the curious thing about Bose's work was that it did not include any vestige of a description of electromagnetic radiation in terms of waves, or indeed of electromagnetism. He had arrived at Planck's equation by treating the photons that fill a cavity as a gas of particles, obeying a different kind of statistical law from the kind of statistics used in the everyday world. The paper had, Bose told Einstein, been submitted to the *Philosophical Magazine*, but had been rejected. He asked Einstein to read the manuscript (it was written in English) and, if he thought it made sense, to pass it on for publication in the *Zeitschrift für Physik*.

Einstein was so impressed by the work that he translated it himself, and submitted it to the journal with his endorsement. Anything endorsed by Einstein was certain of a welcome at the *Zeitschrift*, and the paper duly appeared in print in the summer of 1924. Einstein himself took up the idea of the new statistics, and applied it to other problems; they are now known as the 'Bose–Einstein' statistics.

But how had Bose derived Planck's equation? What were the new statistics?

The simplest way to get a picture of what is going on is to think of a pair of newly minted coins of the same value. If you toss both coins, there are three different outcomes that you might see. There might be two heads showing, or two tails, or one of each. At first sight, you might guess that each outcome is equally probable – that there is a one-in-three chance, for instance, of getting a head-tail combination. But a little thought shows that this is not the case.

Suppose you were to mark one of the coins in some way,

so that the two coins are distinguishable (or use two coins of different denominations). Now, it is easy to see that although there is only one way to get the combination head-head, and one way to get the combination tail-tail, there are *two* ways to get the combination head-tail (think of these as being 'head-tail' and 'tail-head'). So the right way to count the possible results of tossing two coins is as *four* possible outcomes; head-head, tail-tail, head-tail and tail-head. The chance of any one outcome is one in four (a quarter), not one in three. And since there are two ways of getting one head and one tail, the chances of this pattern turning up are one in two, or 50 per cent (a quarter plus a quarter).

Bose found that he could derive Planck's formula by treating photons as particles that have to be counted in a different way. Photons are indistinguishable from one another, and in the photon world the equivalent of this coin-tossing experiment would indeed yield three possible outcomes with equal probabilities. This changes the statistical way in which the photons behave, and the way in which energy is shared out among them – the distribution of photons among different energy states is changed.

There are other features of Bose–Einstein statistics. They apply to particles that are not *conserved*. You make more photons, for example, every time you flick a switch to turn on a light, and they also stream out from the Sun and stars in vast numbers. Photons are constantly being absorbed by the walls of your room, by your eyes, by the surface of the Earth, and so on. But these two processes are not in balance, and the number of photons in the universe is constantly changing.

This is quite different from the behaviour of the kind of particles we are used to thinking of as particles, such as electrons. Electrons cannot be created nor destroyed, except in special circumstances where an electron and its 'antiparticle' counterpart, a positron, are created (or destroyed) together. The total number of electrons in the universe (for this purpose, a positron counts as 'minus one' electrons) stays the same.[4]

The familiar kind of statistics that apply to coin-tossing does, it turns out, apply to conserved particles, such as electrons; these statistics are known to quantum physicists as 'Fermi–Dirac' statistics, in recognition of the work of the Italian Enrico Fermi and the Englishman Paul Dirac. Particles that obey Bose–Einstein statistics, such as photons, are collectively known as 'bosons'; particles that obey Fermi–Dirac statistics, such as electrons, are collectively known as 'fermions'.

The new statistics, nonconservation of photons and indistinguishability of photons are all key components of Bose's work on the nature of a photon gas, although it is not clear that Bose himself fully understood this at the time he sent that paper to Einstein – he had simply found the kind of statistical counting of photons that gave Planck's equation.[5] No matter how he had found the technique, though, the implications were awesome. He had derived the black-body equation for electromagnetic radiation simply by treating photons as real particles obeying a certain kind of statistics and behaving as a quantum gas.

Einstein elaborated on the implications in two scientific papers of his own. Like the exposition of the general theory of relativity, they were read to and published by the Prussian Academy of Sciences. He made the crucial step of appreciating that the argument that light 'waves' could be described purely in particle terms must work both ways – that everyday 'particles' could also be described as waves. In the second of those two papers, published in 1925, he actually said that 'a beam of gas molecules which passes through an aperture must, then, undergo a diffraction analogous to that of a light ray'.[6] By then, however, Einstein had other evidence in support of this extension of wave-particle duality to apply to 'particles' as well as to 'waves'. He referred to the thesis of a Frenchman, Louis de Broglie, who had received his PhD from the University of Paris in November 1924. De Broglie had arrived at the notion of this extension of wave-particle duality by a completely separate route, and is widely remembered today as the originator of the idea that electrons are waves as well as being particles. But even here Einstein played a part, in his still-new role as elder statesman of physics, in getting the idea the attention it deserved.

Particle waves

De Broglie's career had been interrupted by military service during World War I, which is why he did not submit his thesis until the age of thirty-two. While working on his thesis, he published three short papers on wave-particle duality in the journal *Comptes Rendus* in 1923; but de Broglie himself, asked by Abraham Pais in 1978, said that as far as he knew Einstein did not notice these papers at the time.[7] De Broglie completed his thesis at the end of 1923, and early in 1924 he showed it to his supervisor, Paul Langevin, who was so surprised by the revolutionary nature of the work that he asked de Broglie for a second copy of the thesis to send to Einstein for his opinion. Einstein read the thesis, which must have reached him at about the same time as Bose's letter and paper (unfortunately, nobody now knows which one reached him first), and told Langevin that the work was interesting. Once again, a nod from the great man was sufficient. Langevin formally accepted de Broglie's thesis for the award of a PhD on the basis of Einstein's recommendation.

The work that had impressed Einstein in the thesis looks almost too simple, to modern eyes, to have caused so much fuss. In essence, it simply consists of putting together two mathematical expressions that Einstein himself was very familiar with, and had derived for light quantity – the energy relation $E = hf$ (originally due to Planck), and the momentum equation $p = E/c$ (physicists use the letter 'p' for momentum because 'm' is already used for mass). The rather simple new combination that de Broglie derived from these equations is $p = hf/c$. But it is a sign of the revolutionary implications of writing down this equation that Einstein himself had never put the pieces together like this, but had always, before 1924, dealt in his work on light *either* in terms of frequency or in terms of momentum.

De Broglie's equation can be made a little simpler still. Since wavelength is related to frequency by the equation $\lambda = c/f$ (physicists generally use the Greek letter λ [lambda] to denote wavelength), we can write $p \lambda = h$. In plain English, momentum multiplied by wavelength gives Planck's constant. The smaller the wavelength, the bigger the momentum; the bigger the wavelength, the smaller the momentum.

De Broglie derived his equation from Planck's and Einstein's quantum equations of light. Obviously, the relation described the nature of photons. But all moving particles have momentum, and one of the key features of Bohr's model of the atom was that the electrons had to be moving, in orbit around the central nucleus. Why, de Broglie asked, couldn't the wavelength/momentum relation be applied to *anything* that has momentum? Why not use the equation he had found to predict the wavelength of, say, an electron?

The idea became a central feature of the complete new theory of the particle world, quantum mechanics, that was developed in a burst of activity by a handful of scientists over the next few years.[8] At first, de Broglie (and others) believed that every electron must be accompanied by a 'pilot wave', given by his equation, which in a sense told the electron where to go, while the particle itself existed as a separate entity that carried the momentum. But it soon became clear that this is not the case, and that electrons, like photons and all other entities, are both wave and particle at the same time.

From 1925 onwards, almost as soon as the particle-wave idea began to be applied to electrons, Einstein became increasingly alienated from this work. We shall discuss his opposition to the new quantum theory in Chapter Twelve; but before we leave the story of electron waves, we should explain a little more about the basis of the quantum theory itself.

Beyond common sense

As we mentioned in Chapter Six, in 1897, the English physicist J. J. Thomson found negatively charged particles with a mass 1,837 times smaller than that of the hydrogen atom. He had discovered the electron and identified it as a component of the atom. He received the Nobel Prize for Physics in 1906, for identifying the electron as a particle in its own right.

A quarter of a century later, however, de Broglie said that electrons ought to behave like waves, and within a few years, it became possible to study the way beams of electrons were deflected by the atoms in a crystal lattice. Among several researchers who carried out such studies (essentially a variation on the familiar double-slit experiment) was George Thomson, the son of J.J. These studies showed that electrons are, under the right conditions, diffracted by the crystal lattice (just as Einstein had predicted in 1925, although he had referred to beams of molecules rather than electrons) and produce interference patterns. This means that electrons are waves, and George Thomson shared the Nobel Prize for Physics in 1937 with the American Clinton Davisson. J.J., who had received a Nobel prize for proving that electrons are *particles*, thus had the satisfaction of seeing his son receive the prize for proving that electrons are *waves*. Both awards were fully merited; father and son were both correct. Electrons behave like particles; electrons behave like waves.

At the end of the 1980s, Japanese researchers carried out the definitive version of Thomas Young's experiment with two holes, which contains what Richard Feynman called the 'central mystery' of quantum physics. In the standard double-slit experiment for light, a beam of light from a single source is passed through two slits in a screen, and on to another screen. Light waves travelling to a point on the second screen by the two routes travel a different number of wavelengths before reaching it. Where the waves are in step, they add together to produce a bright stripe on the screen; where they are out of step, they cancel to leave a dark stripe. The stripes are interference fringes.

The Japanese team did the same thing by shooting electrons, one at a time, through an instrument known as an electron biprism, and monitoring the build-up of spots of light caused by the arrival of the electrons at a TV screen. As the image builds up, it forms dark and light stripes – interference fringes. Each individual electron behaves like a particle when it strikes the screen, arriving at one unique spot. But the position it arrives at seems to have been determined by a wave moving through *both* slits of the apparatus. It is a wave and a particle at the same time.

Every particle which possesses a momentum p really does also have a wavelength λ, and the two really are related by de Broglie's equation $p = h/\lambda$, where h is, once again, Planck's constant. The dual nature of particles and waves is apparent only at the atomic and subatomic levels, not at the human level, because h is so small – 6.63×10^{-34} joule seconds. The duality is important for electrons, because they have a comparably small mass – just over 9×10^{-31} kilograms.

This wave-particle duality is Feynman's 'central mystery' of the quantum world. It is closely related to the concept of quantum uncertainty – that we can *never* know both the position and the momentum of a 'particle' with absolute precision at the same time.

In the everyday world, a wave is a spread-out thing. The ripples on a pond spread over a long distance, and it is hard to tell exactly where the string of ripples – the wave train – begins and ends. But a particle is a very well-defined thing, which occupies a definite place at a definite time. How can these two conflicting images be reconciled, as they must be if an electron (or a photon) is to be regarded as both wave and particle at the same time?

The appropriate image is of a little package of waves, a short wave train which extends over only a small distance, roughly corresponding to the size of the equivalent particle. Such wave packets are easy to describe mathematically. But the only way to create a wave packet that is localised in space is to allow many waves of different wavelength to interfere with one another. The smaller the wave packet, the more variety of waves with different wavelengths is needed to keep it tightly confined.

This spread of wavelengths corresponds to a spread in momentum, since each unique wavelength has its own specific momentum associated with it, in line with de Broglie's equation. So the *more* precisely the position of the wave packet (= particle) is defined, the *less* precisely its momentum is specified.

You can either know where a particle is, or where it is going, but not both at the same time. If we measure the momentum of an electron, say, precisely, then in a sense we are releasing it from the wave packet and selecting a single wavelength for it. That single wave with a pure frequency extends, in principle, to infinity, so the electron then has *no* unique position. But if we measure its position, we force it into a multi-wavelength state with uncertain momentum. The very nature of reality depends, at this level, on the kind of measurements we make.

Some people mistakenly think that quantum uncertainty is simply an indication of the practical difficulty of measuring small things like electrons. Even today, uncertainty is still sometimes taught (incorrectly) in terms of the way such measurements might be carried out. In order to observe an electron, the argument runs, we would have to bounce radiation off it, and the very act of prodding it in this way will change its position and momentum. That is true, but misses the point. Werner Heisenberg, the German researcher who first appreciated the importance of quantum uncertainty, showed that the uncertainty is a fundamental feature of the nature of an electron or other 'particle'. In the quantum world, objects *do not possess* separate properties known as momentum and position; they carry a mixture of the two, a mixture which can never be completely unravelled *in principle*, not just because of experimental limitations. Momentum and position, and the very idea of a particle, are derived from our experience of the macroscopic world. They simply do not work on the microscopic scale.

The triumph of the quantum

All this led Niels Bohr (taking on board the work of the German Max Born) to develop, in the late 1920s, what is still the standard 'explanation' of the quantum world. Since Bohr had by then returned to Denmark and founded a new research institute in Copenhagen, it is called the 'Copenhagen interpretation' in his honour. The key features can best be understood in terms of what happens when a scientist makes an experimental observation.

First, we have to accept that the very act of observing a thing changes it. We are part of any quantum experiment, and there is no clockwork that ticks away behind the scenes in the same way whether we look or not. Secondly, all we can ever know is the results of experiments. We can look at an electron and find it in position A; then we look again and find it in position B.

We guess that it moved from A to B, but we can say nothing at all about how it did so, and what it was doing while we were not looking at it.

What we learn from experiments is that there is a definite probability that if we look once and get answer A, then the next time we look we will get answer B (and a corresponding, different probability for answers C, D, E . . .). Because there are so many electrons (for example) in everyday systems such as a TV set, the probabilities can be applied with great confidence. Out of each million electrons tweaked in a certain way by an electromagnetic field, a definite proportion, in line with the probabilities, will head off in a certain direction. As long as enough electrons, in a predictable way, travel to the right spot on the TV screen, we don't care how they got there, or what happens to the proportion that end up somewhere else. But underlying such practical considerations, quantum physics deals *only* with probabilities, not with certainties – the discovery that led Einstein, in disgust, to disown it, commenting: 'I cannot believe that God plays dice.'[9] The bizarre features of quantum physics that Einstein found so objectionable are clearly seen not in the workings of a TV set but in cases where the probabilities are more evenly distributed, and we shall describe the classic example of this, the mystery of Schrödinger's cat, in Chapter Twelve.

Einstein didn't believe the Copenhagen interpretation could possibly be an accurate portrayal of the way nature worked; even Richard Feynman never claimed to understand the workings of the quantum world. But every test that we have been able to apply tells us that at the subatomic level particles and waves are two aspects of a single reality, that the outcome of any interaction depends on chance, and that the way we measure things determines the answers we get. There is no clockwork inexorably guiding the workings of the universe from the big bang to the end of time.

The arrival of the new quantum theory marked the end of Einstein's major contributions to physics. He had now passed his own forty-fifth birthday, and very few scientists (or other thinkers) make major new contributions in their late forties and beyond. Almost all the great achievements have been made by researchers in their twenties or thirties. The astonishing thing is not that Einstein's great period of activity came to an end in 1925, but that it had lasted for so long, covering exactly twenty years since the *annus mirabilis*, and that his genius had spread so wide, ranging across thermodynamics, relativity, the nature of light, atomic theory and quantum physics. Einstein still had a major role to play, both in the scientific community and in the world at large. But that role was no longer to be as the originator of breathtaking new insights into the nature of the universe we inhabit.

Chapter Eleven

Exiled from Europe

The year 1923 saw France invading the Ruhr and pushing Germany into an uncontrollable economic slide. The result was near starvation for millions of ordinary Germans who, because of devastating levels of inflation, suddenly found their savings were worthless.

In this atmosphere of desperation, Adolf Hitler, ten years Einstein's junior, found that he had an audience. The Nazis came out of smoke-filled back rooms and into the full glare of public awareness. Before the year was out, Hitler had been arrested and imprisoned for his involvement in the beer-hall putsch, but the world had been introduced to a new group of right-wing extremists – the Brown Shirts.

It was into this politically tense environment that Einstein returned to Berlin and resumed his former pattern of lecturing at the university, attending conferences and travelling abroad to deliver lectures at many of the great European centres of learning.

Although many Germans were ruined by the financial crash of 1923, there were, as always, those who rode out the worst of the crisis and indeed prospered as the country's economy collapsed. The fact that Einstein was financially supported by a number of industrialists who had made their money from the application of science was no secret. Unlike a considerable number of his colleagues, he did not suffer personally or professionally from the ravages of Germany's economic catastrophe. A group of close, wealthy friends placed a considerable lump sum into a bank account from which Einstein could draw out money whenever he wished. This he used to pay for his assistants' salaries and for his own modest material requirements and the account was constantly topped up, even during the worst of the economic troubles of the early 1920s.

In addition, Einstein had made a considerable amount of money out of the success of relativity. This came primarily from his popular account of the subject, *Relativity: The Special and the General Theory*, which by 1920 had run through fourteen German editions and had sold over 65,000 copies – a blockbuster. He also received salaries as a member of the Prussian Academy, titular head of the Kaiser Wilhelm Institute and professor at the university, although, by 1924, the salaries of university professors were slashed to under a third of their prewar levels. Thanks to his international fame and his enormous academic credibility, therefore, Einstein was able to ride out the storm.

Upon his return to Germany, Einstein wasted no time in throwing himself back into the political arena. However, in several cases, he found himself entangled in ineffectual and sometimes embarrassing political situations. One of the clearest examples of this centred on his involvement with an organisation called the League of Nations Committee on Intellectual Cooperation.

The committee, which can be seen as a forerunner of UNESCO, was the creation of the French philosopher Henri Bergson, who saw it as an embodiment of the spirit of an international intellectual union whose main function would be to engender cultural development and education throughout the world. Despite the fact that Germany was not admitted to the League of Nations until 1926, Einstein had been invited to join the CIC in May 1922 as 'a representative of German science'. Without fully questioning the role the committee was to play, he had readily accepted the offer. However, owing to his commitments to Weizmann and his trips to England and the Far East, he was absent from the first meetings of the organisation.

It is quite evident that from an early stage, Einstein was not totally convinced of the usefulness of the CIC. He believed that its members had good intentions but no real power to bring about change. Within two months of agreeing to join, he wrote to Pierre Comert, head of the Information Secretariat at the League of Nations, to offer his resignation.

The response of the organisation was understandable frustration. They saw Einstein as a valuable asset, and felt offended that he should take such a dismissive stance without having played any role himself. Madame Curie, to whom he had written explaining the reasons for his resignation, did her best to persuade him not to resign:

> I have received your letter which has caused me a great disappointment. It seems to me that the reason you give for your abstention is not convincing. It is precisely because dangerous and prejudicial currents of opinion do exist that it is necessary to fight them, and you are able to exercise, to this extent, an excellent influence, if only by your personal reputation . . .[1]

The committee asked Einstein to reconsider. However, by the time he returned to Europe in the spring of 1923, his mind was made up. Breaking his return journey in Switzerland, he took the curious step of making his resignation public by allowing a Swiss paper, the *Nouvelle Gazette de Zurich*, to publish a second letter of resignation even before it had been received by the CIC.

It is unclear how this unfortunate incident occurred. It would not have been in Einstein's nature deliberately to embarass such an organisation as the CIC.

His resignation was rash and it is quite clear that Einstein did not foresee the unfortunate consequences of his move and the fact that such actions merely gave ammunition to his enemies and to the enemies of the committee. For a man of such great intellectual skill and scientific genius, he could on occasion behave in a manner as ill-considered as the rest of us.

Despite Einstein's insensitivity, the CIC still did its best to keep him. However, their efforts were wasted. Einstein had made up his mind and that, at least for the moment, was that.

A few months later, Einstein began to regret the way he had treated his friends on the committee. Writing to Madame Curie in December 1923, he apologised for the fact that he had clearly annoyed her by his resignation and noted that he had not done so because of any personal feelings, but because he felt that the CIC was under political control and therefore could not exert any real influence. He believed that certain individuals were using it for nationalistic aims.

By the spring of 1924, Einstein's feelings towards the CIC had mellowed. When he mentioned to a friend that he regretted the move he had made, the news was passed on in confidence to the secretary of the committe, and by May of that year Einstein had received a letter asking if he would consider reversing his decision of 1923.

Einstein felt that he should accept the committee's offer and, for their part, the CIC knew that they could only benefit from his support. The proposal that Einstein would again join the organisation was placed before the twenty-ninth session of the CIC in June 1924, and unanimously agreed to. Einstein was reintroduced in July and remained a member until 1932, when he resigned for the final time, having decided that the work of the CIC required of him too many compromises for his contribution to be anything other than trivial.

Einstein never felt at home with the Committee on Intellectual Cooperation and it is fair to say that he did very little for it. This was largely due to the powerful political forces at work within the organisation which tied even the most simple proposal in a web of bureaucracy and red tape. This was a system within which Einstein found very little scope to use his particular talents. 'Despite its illustrious membership,' he later wrote of the committee, 'it was the most ineffectual enterprise with which I have been associated.'[2]

On the issue of pacifism, Einstein's stance was quite unbending. He was scathing in his attacks on those who believed that recourse to arms could be justified by political aims. He did not believe in nations and nationalism, and he had a clear vision of a better, more tolerant world, relieved of national borders and international bickering. He strongly believed in the idea of a single world government which, by implication, would mean the abolition of armed forces and would enable the authorities to ensure that all citizens' material needs were provided for. In

Einstein's imagined world, there would be no starvation, no war, no haves and have-nots.

During the course of the next ten years he committed himself to numerous peace campaigns launched by all shades of the liberal political spectrum, attending rallies in countries outside Germany and writing articles and pamphlets stating his views on the issues of international peace and disarmament. He made regular contributions to such publications as the *New York Times*, the US magazine the *Nation* and the *Bulletin of Atomic Scientists*.

Organisers of activist groups and lobbyists for an enormous range of issues in Europe and the United States competed to secure the support of the great man. And Einstein invariably did lend his support. Often he would agree to give a talk or write an article, lend his name to a proposal or petition without knowing the full details. He contributed to such disparate organisations as the Progressive Education Association, the Women's International League for Peace, the Society of Polish Jews and the National Labour Committee for Palestine. He was invited to deliver speeches at Nobel anniversary dinners and award ceremonies at New York's Carnegie Hall, to offer a message of support at the Peace Congress of Intellectuals and to place a message in the time capsule at the 1939 World's Fair.

On many occasions, Einstein and his wife played host to visitors who sought his advice. The strays of the political and scientific worlds beat a path to his door and he was always there with a kindly word or an offer of help. It is said that at one point, hundreds of letters of recommendation signed by the great Albert Einstein were circulating around the academic centres of the world. After a while they of course lost much of their value as it became evident to admissions tutors that he had, in most cases, been cajoled into adding his name to a letter brought to him by one mediocre student after another.

Einstein could never have spared the time to do a fraction of the things he promised, and it was only through the intervention of Elsa that he managed to extricate himself from a plethora of embarrassing situations. As the years passed, Elsa, with her down-to-earth common sense and balanced perspective, took on the role of her husband's business manager and protected him from the pressures that came with his international fame.

Einstein's increased political activity opened further the fissure between him and the establishment. Philip Lenard continued to be an outspoken opponent, who would later embrace the mindlessness of Nazism and turn on all those who wavered over the great political issues of the 1930s. With the growth of anti-Semitism and the expanding political influence of the Nazis, Einstein knew that his days in Berlin were numbered.

Of course, Einstein had many allies, such as Romain Rolland in Switzerland and the leader of the pacifist movement in England, Harold Bing, but they were mainly foreigners and Germans living in other European countries rather than at the epicentre of the political turmoil.

The English philosopher and pacifist Bertrand Russell was a close associate. They not only shared their devotion to logic and science but held almost identical political views. During World War I, Russell had been imprisoned in Britain for pacifist agitation, and he continued to campaign for disarmament until his death at the age of ninety-eight in 1970. He was an early supporter of relativity and published a highly successful book on the subject in 1925 called *The ABC of Relativity*. The book was much admired by Einstein and served to forge a friendship between the two men which was to continue for the rest of Einstein's life.

In a serialised form of his book that appeared in the *Nation* during 1925, Russell expressed the belief that once people had become used to the idea of relativity it would change the way they thought: people would work with greater abstraction and would replace old absolute laws with relative concepts. This has certainly happened in the world of science but the absorption of relativity into popular culture has done little to change the way most people think, simply because very few have got used to relativity or understand it in the least.

As the foundations of the quantum theory were established during the first half of the 1920s, Einstein settled back into the pattern of life before the great events of 1919.

The family continued to live at Haberlandstrasse 5, a large flat on the fourth floor of a six-storey building constructed in traditional style with two balconies bedecked with troughs of flowers and fronted with clusters of rectangular windows overlooking the busy street.

It was a comfortable home, perfectly suited for a husband, wife and two daughters. It was decorated in the fashion of the time – parquet flooring covered with colourful rugs, the wallpaper elaborate, the furniture dark and heavy. A grand piano took pride of place in one of the living rooms and Einstein frequently played or accompanied others on the violin.

The home was completely dominated by Elsa. She chose the fabrics and the colour schemes, furniture and ornaments. Einstein handed all such responsibilities to his wife. The only room in the house which was sacrosanct was Einstein's study. Nobody was allowed in there without an invitation and Elsa was strictly forbidden to clean or dust there. Einstein's office was in a perpetually chaotic state, his desk piled high with books and papers, the overloaded shelves dusty with neglect, and the floor visible only in places where gaps remained between piles of books and papers. Yet he always knew exactly where a particular paper or book was kept. When he was not at the university or travelling to lecture abroad, Einstein would spend the entire day there working at his desk, filling page after page with closely packed equations written in his neat, straight handwriting.

Einstein and Elsa were in constant demand on the Berlin social scene and within the higher echelons of the German academic world. The couple were frequent guests at gala dinners and formal university functions. Einstein had a rebellious approach to such events, seeing them as pretentious and time-wasting. He would often appear at a formal function wearing shoes but no socks, a habit which became one of his hallmarks as tales of his eccentricities spread through the international media.

Thus we have the story that Einstein for months used a cheque for a considerable sum as a bookmark until it was noticed by a colleague. He refused to buy tails for a formal occasion and only after being harangued by Elsa did he compromise and agree to the purchase of a dinner jacket. When asked by a Berlin socialite if he would attend a celebrity dinner, he replied: 'So, you would like me to serve as a centrepiece?' On another occasion he remarked that such events were akin to 'feeding time at the zoo'.

At home, Einstein could relax. He could wander the apartment in bare feet, play his violin and, thanks to the filter system created by Elsa and her daughters, he could pore over his books and papers without being disturbed.

The Einsteins were a close family and Elsa's daughters Ilsa and Margot adored their stepfather. Ilsa was married at the age of twenty-seven in 1924, leaving Margot, the younger by two years, with Elsa and Einstein. Both daughters were to keep close links with their mother and stepfather for the rest of their lives.

Within his family, and with a few of his closest friends, Einstein could escape from the adulation of the world – an especially precious retreat when the glare of public acclaim was so bright. At home there were no pretensions. Although they respected Albert's study as an inner sanctum, to Ilsa and Margot, Albert Einstein was simply their father.

Before Ilsa left home, the family had a visit from a foreign artist who came to sculpt Einstein. Being only vaguely familiar with the German language, she insisted on referring to her model as *der Genie* ('the genius'), which caused much amusement in the Einstein household, where Albert was henceforth known as the Genius: When would the Genius be home for supper? Did the Genius have a good nap? From then on, when a guest arrived for Einstein, he or she would be announced as a visitor for the Genius and be confronted by the laughter of the three women of the house as well as smiles from Einstein himself.

Einstein had been weakened by the succession of serious ailments he had faced towards the end of World War I, and in February 1928 he fell seriously ill while on a visit to Davos in Switzerland.

Arriving late one evening in the town of Zuoz on his way back to Germany, he had decided to stay with a friend, Willy Meinhardt. Not expecting Einstein to reach the station when he did, Meinhardt was not there to meet him. Typically declining the help of a porter to carry his heavy bags, Einstein set off walking in the freezing cold. Before going more than a hundred yards he collapsed and was rushed to hospital.

Einstein was diagnosed as suffering from inflamed walls of the heart and was put on a special salt-free diet. It was suspected that one of the causes of the trouble was the fact that he did very little physical exercise other than walking, and that, if anything, he was too well cared for by Elsa. Back in Berlin, Einstein was confined to bed and, when duties permitted, his doctor, the family friend Janos Plesch, saw to it that he spent some time with Elsa and her two daughters at a seaside resort on the Baltic coast, where he could relax and fully recover.

It was through this unexpected illness that a key figure entered Einstein's life. By April it had become clear that he simply could not cope with all his correspondence. Every day he received mail pertaining to his scientific work, his involvement in various peace movements and Zionist efforts, invitations to lecture at a variety of academic institutions around the world, and personal letters. Elsa could not take on the responsibility for all this as well as running a home and looking after a now incapacitated husband. Einstein hired a secretary named Helen Dukas, who was to remain his personal assistant until his death twenty-seven years later.

Helen Dukas came from Swabia. She was an intelligent woman but had no scientific training. She had boundless energy and was hyper-efficient. During the following years, she became an indispensable help to Einstein and, with Elsa, she protected him from the outside world to the best of her ability. She would invariably travel with the Einsteins on lecture tours and when the family moved to the United States in 1933, she moved with them and adopted the role of housekeeper-cum-secretary after Elsa's death in 1936.

Einstein's recovery was slow and he did not regain his health for the best part of a year. During that time, the world political and economic situation had changed once again. In 1929, the money markets of the world were devastated by the Wall Street crash and the chaos it engendered reverberated around the globe. Once again Germany was hit by recession and further economic catastrophe.

Thanks to Einstein's social standing, international fame and the protection of friends and employers, the Einsteins managed to ride out this second wave of economic trouble. By this stage in his career, Einstein's financial situation was secure. The greatest threat to his wellbeing came not from the economic uncertainties of the time but the political ferment brewing just beneath the surface of the German social structure.

Aside from the economic problems of the country, 1929 was a time of celebration for the Einstein family. Albert's fiftieth birthday on 14 March was an event celebrated around the world and by millions of complete strangers as much as by the intimate circle of his immediate family. Einstein had anticipated the overreaction of the media, and the family escaped to the house of their friend Dr Plesch at Gatow.

In the event, the day turned out to be a blend of great warmth and genuine feeling from an admiring public combined with a hint of the absurd.

The day began with literally hundreds of presents arriving at the empty apartment on Haberlandstrasse. These included an ounce of tobacco sent by an unemployed labourer who enclosed a note stating that it was 'a relatively small amount but gathered in a good field'. Knowing that one of his greatest pleasures was sailing, several of his close friends clubbed together to buy Einstein a sailing boat, which he named the *Tummler*. It brought him many hours of happy relaxation.

By coincidence, along with all the telegrams and messages of good will from around the world arriving that day, there came a less than welcome visit from a minor tax official. The tax inspector had learned that Einstein was staying in Gatow and had come to discuss his tax return. It was only when the two men had sat down to sort out the matter that Elsa informed the inspector that it was the great scientist's birthday. Deeply embarrassed by his heavy-handedness, the visitor offered his sincerest apologies and congratulations and beat a hasty retreat.

Einstein could not understand the fuss being made over the fact that he had reached fifty, but he was genuinely grateful to his friends, colleagues and the legions of strangers around the world who had given him presents, sent him messages and displayed such affection. To many, he sent a mimeographed copy of his own doggerel verse, accompanied by a personal letter. The verse read:

> Everyone shows their best face today,
> And from near and far have sweetly written,
> Showering me with all things one could wish for
> That still matter to an old man.
> Everyone approaches with nice voices
> In order to make a better day of it,
> And even the innumerable spongers have paid their tribute.
> And so I feel lifted up like a noble eagle.
> Now the day nears its close and I send you my compliments.
> Everything that you did was good, and the sun smiles.

However, not every aspect of Einstein's birthday went quite as smoothly as he suggested in this verse.

In keeping with the international good will expressed on the occasion of Einstein's fiftieth birthday, the Berlin city authorities decided to give Einstein a present to acknowledge his invaluable contribution to science. They had been prompted by Janos Plesch, who, early in 1929, had approached them to try to persuade them to make a gesture on the forthcoming occasion. According to legend, the mayor of Berlin did not know who Einstein was or what he had done, and it took all Dr Plesch's powers of persuasion to convince them of the merit of bestowing a gift on the internationally renowned scientist who lived in their city.

In the event, a gift was made of a house with an accompanying piece of land situated near the river Havel. It was widely known that Einstein loved sailing and the site had been chosen with this in mind. The local media made much of the present, doubtless surreptitiously informed of the fact by the publicity-conscious city council. But when Elsa went to view the house, she found that it was already occupied and that the tenants had no intention of leaving.

The council appeared to be shocked by the news and with profuse apologies offered another plot of land near the original site. The only problem with this lay in the fact that when first leased it was under the condition that nothing could be built on the land. Thus the Einsteins were to be presented with a wholly useless piece of real estate. It is unclear what the actions and motives of the council were during this scenario, but there were anti-Einstein elements within the council, and the second plot of land was given with the full knowledge that nothing could be built on it.

It was some time before a solution of sorts was worked out. Having by now destroyed all good will surrounding the gift, the council finally said that the Einsteins could choose a plot of land for themselves. Albert and Elsa selected a site near the village of Caputh a few miles from Potsdam. The council agreed to buy the land for the Einsteins, who then funded the building of a house to their design and specifications.

The house in Caputh severely drained the family's financial resources at a time when money was a problem for everyone, but they did come out of the affair with a beautiful home in which they spent as much time as they could away from the hustle and bustle of Berlin. However, even as early as 1929 Einstein realised that the family's time in Germany was limited and that the beauty and tranquillity of the house in Caputh would be a transitory pleasure. The onslaught of fascism and the ever-worsening anti-Semitic feeling building up in the country now seemed unstoppable.

By the time the house in Caputh had been completed, the Einstein's were to remain in Germany for little more than three years, and much of that was spent out of the country.

During the course of the next three years, Einstein's commitments fell into a neat pattern where he could divide his time between several centres. This enabled him to spent a good deal of time away from the increasingly oppressive atmosphere of Berlin without having to resign from his positions there.

Through his contact with the American physicist Robert Millikan, Einstein was invited to the California Institute of Technology in Pasadena. Although he did not share Einstein's political views, Millikan was a great admirer of his scientific work and saw that even the temporary acquisition of his talents would prove a significant catch for Caltech. Millikan was supported in this by the other scientists based in Pasadena, but did encounter opposition from non-scientists who were concerned by what they saw as Einstein's extremist stance on pacifism and the structure of international politics. Millikan was so keen to have Einstein at Caltech that he put aside his own opposition to Einstein's political activity and defended his colleague on many occasions, including, on the eve of one visit, a venomous attack from a Major General Amos A. Fried, who claimed that Einstein would be teaching treason to the youth of America.

Einstein was given the title of visiting professor at Caltech and first took up the post in December 1930. In the event, the three months' stay was a mixture of work and play. As well as working at Caltech, he came into close contact with the team, headed by the astronomer Edwin Hubble, who worked at the 100-inch telescope at Mount Wilson, near Pasadena. At the time the instrument was the most powerful telescope in the world and it was through the efforts of Hubble's team that experimental support for early cosmological theories had been developed during the 1920s.

The Einsteins were the centre of attention during their brief visit and were again courted by the rich and famous. They visited Hollywood and, after it was discovered that the two men enjoyed a mutual respect, a meeting was arranged with one of the greatest stars of the time, Charlie Chaplin.

Einstein was the first scientific star of the media age. Newton and Galileo, Leonardo da Vinci and Aristotle were all famous in their own countries and within the confines of an age when mass communication was limited. However, with the advent of radio and the global distribution of newspapers, and road and rail transport, the world had opened up. People in all walks of life and in every continent could share in the phenomena of relativity and its charismatic creator.

The Einsteins returned to Berlin in mid-March 1931 and spent the spring in Caputh. Einstein occasionally travelled into Berlin to work at the university, but spent most of his time working in the study of his country retreat. However, the stay in Germany was short-lived; by May he and Elsa were abroad again. This time their destination was Oxford.

It had been almost four years earlier, in 1927, that Einstein had first been approached to lecture at Oxford. The invitation had come from his friend and colleague Frederick Lindemann, who had been Einstein's host during his brief visit to the city in 1921. It was Lindemann's intention that Einstein should be awarded an honorary degree and deliver a series of lectures funded by the recently established Rhodes Trust, at Rhodes House.

Initially, Einstein had turned down the invitation to Oxford because of overcommitments in Berlin. Then his protracted illness and convalescence intervened, and it was not until he had fully recovered that the visit could be arranged.

Einstein's first Rhodes lecture, in May 1931, was not at all successful, for the sole reason that it was delivered in German and understood by only a fraction of the audience. The majority of those who had been listening intently at the beginning had left before the end of the first half-hour of the talk. Henceforth, the lectures were delivered in English and received general acclaim.

Einstein loved Oxford. It offered him the tranquillity he craved and had found in the smaller cities of Germany and Switzerland. He found the pace of life in Oxford perfectly conducive to work and thought.

By all accounts, Einstein was treated with a blend of affection and respect by those around him. Anecdotes of the time he spent in the city abound. One which rings true to his character is the story related by a porter at Christ Church College.

As Einstein was disembarking from a car he lost a button from his coat. A young female student among the group picked it up and ran after Einstein to return it. At this point she was stopped by the porter, who claimed that she was wasting her time because the gentleman already had an odd button on his coat and wouldn't notice the fact that he had lost one.

Lindemann was keen to secure a permanent position for Einstein at Oxford. He realised that he could probably never induce the scientist to move to the city permanently, but tried for the next best thing – a research fellowship at Christ Church College, the most prestigious honorific position the college could offer.

It proved more difficult than Lindemann had imagined. There existed at Christ Church a large and vocal group of scholars, mainly classicists, who perceived scientists as being 'unclubbable' and therefore a threat to the social life of the college. To complicate matters, some English academics of the time were not above anti-Semitism. One member of the college went as far as to announce that he did not believe that the college funds had been given to subsidise 'some German Jew'.

Soon after Lindemann's proposal had been put forward, the internal college battle lines were drawn up between a large group who wanted Einstein at the college and a smaller but influential set who did not approve.

In the event, Einstein's acceptance into Christ Church had little to do with the efforts of Lindemann and more to do with poetry. During Einstein's first visit to Oxford, he had been given the rooms of one R. H. Dundas, who was abroad. Dundas exerted a great deal of influence at Christ Church and was a close friend of the dean. Before leaving the college, Einstein left a thank-you note for Dundas, written in verse. When he returned to Oxford and found the note, Dundas was so delighted by Einstein's poem that he voiced his support for Einstein's research fellowship and the motion was passed.

The position at Christ Church was granted for five years with an annual payment of £400. According to the conditions of the position, Einstein had only to spend a short period of each year in Oxford. He would be provided with a college room and dinner allowance when he decided to dine in Hall.

With the offer from Oxford, Einstein managed to establish a new work pattern and one which deliberately involved spending very little time at the University of Berlin. During the course of the next two and a half years, he made two more visits to Pasadena and several to Oxford, and when he was based in Germany he spent most of his time working at home.

The house at Caputh was idyllic. It was situated in the woods near the village and constructed of logs, open-plan, with large windows on all sides. In one direction it commanded splendid views of the Havelsee, where Einstein loved to sail his boat; in the opposite direction lay the forest and beyond that the red roofs of Caputh. It was a quiet and peaceful place to work and because Einstein had no teaching commitments at the university he was free to spend as much time as he wished in the beautiful surroundings of the countryside he loved.

By travelling or retreating to the country, the Einsteins could withdraw from the increasingly ugly political mood in Germany. Einstein was fully aware of the deteriorating situation and found the atmosphere within the country a little more sour upon each fresh return from abroad. By the turn of the decade there was little doubt in his mind that he and his family would have to leave.

As Einstein sat in his home in Caputh struggling with his work on the unified field, the seeds for his future life were being planted across the world in the United States.

In 1929, the American educationalist Abraham Flexner had acquired funding for the construction of an advanced research institute in Princeton, New Jersey. The money, estimated at $5 million, had been pledged by a couple of New Jersey business tycoons, Louis Bamberger and his sister Caroline Fuld.

The idea was to create a centre for the advanced study of science where eminent figures from the international scientific community could live and work in a peaceful and productive environment, free of any lecturing responsibilities. It was evident that if this institute could be set up and made attractive enough to entice the big names of the scientific world, then, as a long-term investment, it would produce results from pure research that could find practical application. In essence, Flexner's concept was identical to the establishment of research institutes in Germany under the control of Kaiser Wilhelm some twenty years earlier.

By 1931, Flexner, as yet without a site for his institute, was in the process of head-hunting important scientists from Europe and the United States to work at his establishment. Early in 1932, he decided to visit Pasadena to seek the advice of Millikan.

Little did Millikan realise it at the time, but during the course of one brief conversation with Flexner he effectively set in motion a set of circumstances which would eventually lead Einstein to Princeton and out of Nazi Germany. After telling him that he could be of little real help, Millikan suggested that Flexner should have a word with Einstein.

Einstein was in Pasadena on his second annual visit to Caltech. According to later accounts, he was interested in what Flexner had to tell him of his plans; for his part, Flexner had not the slightest thought of actually inviting Einstein to join his institute; a story one may or may not take at face value. Whatever Flexner's thoughts at the time, he arranged to meet with Einstein again later in the year.

The two men met up again in Oxford in the late spring of 1932. Einstein was there for his first term as a research fellow. Again, nothing conclusive came from their discussions, but Flexner told Einstein that, although he would never presume to invite him to join the institute, if he ever decided that such an establishment would suit his needs, then he could join on his own terms. Before parting, the two men agreed to meet again when Flexner was to visit Germany in the summer.

Within two months of their meeting in Oxford, Flexner was staying with the Einsteins at Caputh. During the American's short visit, he and Einstein went on many long walks in the summer sunshine, deep in conversation as they skirted the lake's edge and trod the well-worn paths through the forest around the wooden house.

During the course of these conversations, Einstein's interest in joining the project grew. What Flexner was offering was a tempting addition to Einstein's overseas commitments at a time when the situation at home was looking bleak. By the end of Flexner's visit, Einstein had agreed to join the Institute of Advanced Study from the autumn of 1933, based on the idea that he would divide his time equally between Berlin and Princeton. The only remaining problem was obtaining the agreement of the university authorities in Berlin. Flexner went away a very happy man and Einstein set about arranging things with the university.

One story typifies Einstein's approach to discussions on the subject of money. Asked what he would wish to be paid as a member of the institute, Einstein thought for a while and then said; '$3,000 a year? Could I live on less?' To which Flexner replied: 'You couldn't live on that', and suggested that the matter would best be sorted out with Mrs Einstein. Shortly afterwards, Flexner and Elsa arrived at a figure of $16,000 per annum, to be continued after retirement.

Even as early as 1932, Flexner was probably thinking in terms of ensnaring Einstein on a permanent basis. Flexner realised that if Einstein's first six-month stay in Princeton during the autumn of 1933 was to the scientist's liking, with the growing difficulties in Germany, he might well be persuaded to stay.

What is certain is that by attracting Einstein to the institute, Flexner had pulled off quite a coup and by a single stroke had enormously enhanced the profile of his project within the scientific community. Flexner had thereby driven a wedge between Einstein and Caltech and had effectively killed off any chance of Millikan doing for Pasadena what Flexner had done for Princeton. Caltech is a highly prestigious institution and a very attractive place to work, but Flexner had, nonetheless, managed to lure Einstein to the east coast.

What is more, Flexner appears to have done all this without Einstein realising what was happening. If Einstein had been aware of Flexner's manipulations he would almost certainly not have gone along with them and might well have lived out the rest of his days on the west coast of America rather than in the east.

With his future plans laid, Einstein spent most of the autumn of 1932 at the house in Caputh, but by December of that year, he and Elsa were once more packing for their annual winter stay in Pasadena. However, this time, Einstein knew that this trip would not fit into the usual pattern of their recent visits. The routine preparations were the same as they had been in the previous two years, but under his show of normality, Einstein knew that their days in Caputh were near their end.

In early December, Albert and Elsa left the house to catch the train to Antwerp at the start of the long journey to the United States. As they stepped out of the front door, Einstein turned to Elsa and said: 'Before you leave our villa this time, take a good look at it.' When she asked why, he replied: 'You will never see it again.'

Elsa did not share her husband's sense of doom and later told a friend that she had thought that, at that moment, Albert was being rather melodramatic.

In the event, Einstein was absolutely right. He and Elsa left German soil on 10 December 1932 and never set foot in the country again. While they spent the winter in California, important events were unfolding in their homeland.

On 30 January 1933, Adolf Hitler came to power in Germany. Anti-Semitism was one of the central themes of his party. The Nazis espoused fascism as the new order through which Germany would once again become a world power. It was not the sort of country in which Einstein could or would want to live and work.

Einstein had expected the worst and the news was received in the Californian sunshine with equanimity. Back home in Berlin, Einstein's bank account was frozen, the house in Caputh raided to search for weapons allegedly hidden there by communists, and copies of his popular book on relativity were burned publicly. It was obvious that Einstein could not return.

By March 1933, Einstein was in New York. There he visited the German consul, who officially declared that there was nothing to stop Einstein from returning to Germany, but privately made it clear that he thought such a move would be highly dangerous. Einstein had already decided what he was going to do.

Declaring that he was to return to Europe, but to Belgium rather than Germany, he held a press conference at the Waldorf Astoria to launch his latest book, *The Fight Against War* and took the opportunity to renounce publicly the National Socialist government and to align himself with the peace movement. The

German press responded in the expected fashion; one newspaper declared: 'Good news from Einstein – he's not coming back.'[3]

It was not only the mass media that, under the strain of political pressure, turned against Einstein. In March 1933, Einstein resigned from the Prussian Academy of Sciences shortly before they were to expel him, and in April of the same year he resigned from the Bavarian Academy of Sciences.

The Prussian Academy had claimed that Einstein had subverted the policies of the German government by disseminating atrocity stories outside the country. When Einstein wrote to the academy denying these charges, they were withdrawn, but the bitterness which had been created by the whole fiasco remained for the rest of Einstein's life.

Although many scientists and other academics were influenced by the mood of the times and swept up by the wave of nationalism which came in the wake of Hitler's rise to power, the actions of the academy were strongly opposed by many of Einstein's loyal friends. Nernst, Planck and the physicist Max von Laue all remained on Einstein's side and it is claimed that during a private interview, Planck stood up to Hitler himself over the matter of Einstein's scientific work.

Planck also came out strongly in Einstein's defence at a plenary session of the Prussian Academy in May 1933, soon after Einstein's resignation:

> I believe that I speak for my Academy colleagues in physics and also for the overwhelming majority of all German physicists when I say: Mr Einstein is not just one among many outstanding physicists; on the contrary, Mr Einstein is the physicist through whose works published by our Academy, physics has experienced a deepening whose significance can be matched only by that of the achievements of Johannes Kepler and Isaac Newton . . .

Einstein did not stop at his resignation from the Prussian and Bavarian academies. For the second and final time in his life he renounced his German citizenship. In a classic exercise in bureaucratic futility, the German government then made a public show of withdrawing Einstein's citizenship. Einstein commented many years later that the action of the German government was akin to the public hanging of Mussolini's dead body after he had been executed.

With the German government and much of the German academic community aligned against him, Einstein set sail for Europe at the end of March 1933. Ahead of him lay over six months of uncertainty and concern for his and Elsa's future in a Europe on the brink of war.

At this point it became clear that Flexner's offer to work at Princeton could be extended to a permanent position if Einstein wished. Flexner was certainly keen to realise such a plan and Einstein, having made a brief visit to Princeton before leaving New York, was increasingly interested in the suggestion. Elsa was comfortable with the idea of living in New Jersey. However, there was also an anti-Einstein movement in the United States.

Although the vast majority of people in America and Europe outside Germany were fond of the famous scientist and fascinated by his work, even in those countries which were, in a few years, to be allies against Germany, there was a body of anti-Semitic and anti-pacifist opinion. Einstein had experienced such feelings at first hand shortly before his last visit to the States, when the National Patriotic Council had issued a statement calling him a Bolshevist and claimed that he espoused 'worthless' theories. During the same week, the American Woman's League had branded him a communist, declaring that the State Department should refuse to give Einstein an entry permit.

To the latter, Einstein publicly replied in tones both humorous and bitter:

> But are they not right, these watchful citizenesses? Why should one open one's doors to a person who devours hard-boiled capitalists with as much appreciation and gusto as the Cretan Minotaur in days gone by devoured luscious Greek maidens, and on top of that is low down enough to reject every sort of war, except the unavoidable war with one's own wife? Therefore give heed to your clever and patriotic womenfolk and remember that the

> Capitol of mighty Rome was once saved by the cackling of its faithful geese.[4]

Whatever their reasons, the Einsteins did not go straight back to the United States. The six months from April to October 1933 were divided between the Belgian seaside resort of Le Coq sur Mer and the dreaming spires of Oxford. This was one of the most insecure and haphazard periods in Einstein's life, a time when he could quite accurately call himself a dispossessed Jew.

Le Coq sur Mer, as well as being a very picturesque spot, was considered to be a comparatively safe retreat for the Einsteins. But they knew that Einstein's life was in real danger, and on Elsa's insistence he had a bodyguard at all times.

During the late spring of 1933, the Nazis produced a catalogue of state enemies which somehow made its way to the Einsteins in Le Coq sur Mer. It contained the most unflattering photographs of the regime's opponents with short captions written underneath. Einstein headed the list and under his photograph was written 'not yet hanged'.

Although the Nazis were keen to keep a clean image abroad, they went to great lengths to discredit Einstein. He was accused of being the leader of the communist movement in Germany and responsible for various arms-smuggling operations. On many occasions, the Nazis tried to involve Einstein in espionage by sending bogus underground agents to visit him in an effort to enlist his help which they could then expose as treason. Naturally, Einstein did not fall for such crude devices.

If it was not for the disastrous state of his home country and the fact that Europe was sliding irretrievably into a state of war, the early summer of 1933, would have been an idyllic time for Einstein. When he was not in Oxford, he found himself living in a comfortable and remote little house on a Belgian beach with Elsa. His stepdaughters visited frequently and under the benign, watchful eyes of a pair of heavy-set bodyguards, Einstein and Elsa played hosts to many visiting friends.

The summer of 1933 was a time of great moral dilemma for Einstein. It was then that he was faced with the problem of how to justify his pacifistic stance in the face of the Nazi threat.

Einstein had always held the view that violence and military action were inexcusable under any circumstances. However, it was already obvious that the Nazis were a great threat to the very fabric of a free Europe. As a consequence, Einstein reached the conclusion that there could be no role for pacifism in the present world situation and that the only way in which the Nazis could be defeated was by fighting. Only in this way, he believed, could the world become a better, safer place. War represented the lesser of the two evils facing civilisation.

Things came to a head towards the end of July. Einstein had been asked by a young French pacifist named Alfred Nahon to speak on behalf of two Belgian conscientious objectors. Fully expecting the great scientist and world-famous pacifist to come down squarely on the side of the objectors, Nahon instead received a letter from Einstein outlining his views and requesting that the letter be made public. The letter read:

> What I shall tell you will greatly surprise you . . . Imagine Belgium occupied by present-day Germany. Things would be far worse than in 1914, and they were bad enough even then. Hence I must tell you candidly: Were I a Belgian, I would not, in the present circumstances, refuse military service, rather, I would enter such service cheerfully in the belief that I should thereby be helping to save European civilisation. This does not mean that I am surrendering the principle for which I have stood heretofore. I hope most sincerely that the time will once more come when refusal of military service will again be an effective method of serving the cause of human progress.[5]

When this letter was published in a French magazine a short time later, shock waves went through the entire pacifist movement. Einstein's views were naturally seen as a U-turn, and hardline pacifists were outraged.

Of course it could be argued that Einstein was reversing deeply held views just when they were about to be put to their first serious test and that his action was indeed a U-turn, if not a complete renouncement of his widely received wisdom of barely weeks previously. Was this the same Einstein who, four years earlier, had stated publicly: '[In the event of War] I would unconditionally refuse all war service, direct or indirect, and I would ask to persuade my friends to adopt the same position, regardless of how I might feel about the causes of any particular war'?[6]

Clearly, Einstein had thought long and hard about what could be done in the face of the evil Germany of 1933. It is evident from the views of friends and colleagues who were with him at the time that the decision in favour of militarism did not come easily and pained him. But he could see no other way.

Although the German government, the anti-Semites and the anti-communists across the globe made much capital out of the controversy, their gains were short-lived. Many intellectuals eventually sided with Einstein's new analysis of the situation. It simply took them a while to come round to his way of thinking. Bertrand Russell saw the logic of Einstein's argument and quickly followed his lead. In Russell's wake many wavering intellectuals and former activists followed.

Einstein made many social visits to the royal family during his time in Belgium in 1933. He had first met King Albert I and Queen Elisabeth in 1929, and found them to be civilised and sophisticated people with a keen interest in intellectual matters. They in turn were fascinated by Einstein's work as well as attracted by his easy-going and relaxed personality. Their conversations ranged across the spectrum of scientific and political matters and also touched on the matter of what Einstein and his wife were planning to do in the future.

There are many anecdotes that illustrate Einstein's eccentric and often forgetful ways, and one of the most amusing concerns his friendship with the Belgian royal family.

In the summer of 1933, during one of his visits to England, Einstein was in London and decided on a whim to visit the king and queen of Belgium. Although he had been given a substantial sum of money by Elsa before he left for England, he had apparently given most of it away and found that he barely had enough money to purchase a third-class ticket. However, he often travelled third class, against Elsa's protestations. He caught the train. Upon arriving in Belgium, he decided to look for some cheap lodgings before travelling on to the queen's summer residence at Laeken. Eventually, Einstein ended up lost and exhausted in a slum district of Brussels. Finding the nearest inn, he asked how one might reach Laeken by telephone.

The barman did not recognise the world-famous scientist. The elderly man standing before him clutching a small bag and a violin case looked like a tramp. His hair was matted and dusty and his clothes hung on him, crumpled and dirty. When the landlord asked 'Why Laeken?', Einstein said; 'I want to reach the Castle of Laeken – Queen Elisabeth.'

The landlord directed Einstein to the telephone kiosk in the corner of the bar and returned to his customers. After a brief discussion, the barman and his customers reached the conclusion that the stranger was a madman. He might be a dangerous lunatic, perhaps an anarchist, plotting to assassinate the royal family? They decided to call the police.

By the time Einstein had worked out how to use the telephone, had managed to get through to a courtier and had re-emerged from the booth, a large group of anxious locals had gathered around the kiosk, headed by two Brussels policemen who proceeded to take Einstein away for questioning. It was only after urgent phone calls had been made that the true identity of the bedraggled stranger was discovered and he was escorted to his royal friends at Laeken.

Another widely reported tale describes how, on at least one occasion, Einstein telephoned Elsa while on his way to an important meeting and asked; 'Where am I and where am I meant to be?'

In England, Einstein spent most of his time working quietly at Christ Church College in Oxford. During these visits he delivered many public lectures and gave talks on pacifism and anti-Semitism. The most important of his political efforts took place on 3 October 1933. This was to be his last large-scale public appearance in Europe and took place shortly before he set sail for the United States. The venue was the 4,000-capacity Royal Albert Hall in London, where he spoke on behalf of the

Academic Assistance Council, an organisation that was trying to help students who, in the present political climate, could not continue their education.

On the occasion of Einstein's talk, the hall was packed to bursting and many hundreds were turned away at the door. The excitement of the evening was further enhanced by a rumour that there would be an attempt on Einstein's life during the course of the evening. The hall was consequently heavily guarded by a 1,000-strong security force composed largely of students supplemented by extra police drafted in for the evening.

In the event, there was no sign of trouble from right-wing activists and the evening went off without a hitch. Einstein delivered a passionate speech demonstrating the importance he placed on the subject of education and the right of every individual to have educational opportunities – a cause for which he had energetically campaigned for many years.

Einstein left Britain shortly after the Albert Hall speech. It is clear from conversations with Lindemann and his other colleagues in Oxford that, even at this stage, he fully intended to return for his annual summer visit. As he boarded the *Westernland* and was reunited with Elsa, who had embarked at Antwerp, it is quite possible that he realised that such plans might not after all become reality. And, as the liner pulled out of Southampton to begin the long voyage across the Atlantic, he may well have looked out across the port and thought that this would be his last view of Europe.

It is of course impossible to know what was going through Einstein's mind as he left England in early October 1933. Back in Germany, the mad dogs of the fascist state had been let loose and were doing their worst. Within a few short years, Europe would be aflame with war, and millions of men, women and children would be dead as a result. Among those lost victims of the conflict would be 6 million Jews, Einstein's own people, his kin. He was one of the lucky ones. Because of his unique talents he was able to escape the horror to come. But, what lay ahead for him in a strange country, where the people spoke a different language and had different customs? All he could really rely upon in the New World was the unifying principles of his work. As far as physics was concerned, he and his new hosts spoke the same tongue, and it was through science that he could build a new future for himself and his family. Ironically, though, just at the time Einstein was exiled from Europe, he also became exiled from the mainstream of science, fighting an unavailing rearguard action against the quantum theory he had helped to found, but had come to distrust.

CHAPTER TWELVE

Quantum Opponent

Louis de Broglie's version of wave-particle duality was taken up and developed by an Austrian physicist, Erwin Schrödinger. In the paper in which he talked of a beam of molecules being diffracted, Einstein had mentioned de Broglie's thesis work, with the comment 'I believe that it involves more than merely an analogy'.[1] The paper was published in February 1925, and it was when he read those words that Schrödinger took up the idea of electron waves, and developed a complete mathematical description of the behaviour of electrons in atoms, that became known as 'wave mechanics'. This was published in 1926, almost exactly at the same time as another complete mathematical description of the behaviour of electrons in atoms, developed largely by Werner Heisenberg, which treated electrons as particles and was known as 'matrix mechanics'. Almost immediately, Paul Dirac showed that the wave and particle treatments were mathematically equivalent to one another, and always give the same 'answers' to problems involving the behaviour of electrons. The new theory became known as quantum mechanics.[2]

In practice, most physicists, to this day, use Schrödinger's version of the equations. This is largely because they are more familiar. Every high-school physicist learns about waves and the equations you need to describe the behaviour of ripples on a pond, or electromagnetic waves, and finds it easy to move on to Schrödinger's equations; matrix theory is not usually taught properly until you get to a higher level of education, and so it is less comfortingly familiar. This is unfortunate, because it leaves even many physicists with the gut feeling that the 'right' way to describe things like electrons is as waves. Schrödinger himself started out with that thought in mind, trying to get rid of the strange notion of electrons 'jumping' between energy levels, and to remove the element of chance in quantum theory that Einstein also found abhorrent.

When it was proved that wave mechanics and matrix mechanics were the same thing dressed in different mathematical clothes, and Niels Bohr and Max Born established an interpretation of quantum mechanics that includes not only wave-particle duality, chance and uncertainty but also the role of the observer in deciding the outcome of experiments, Schrödinger was deeply upset. Disgusted that even his beloved wave mechanics did not get rid of the need for quanta themselves and quantum jumps, he made a remark which is widely quoted (with slight variations in the translation): 'Had I known that we were not going to get rid of this damned quantum jumping, I never would have involved myself in this business.'[3]

Einstein didn't like it

Einstein didn't like it either. But before we look at his objections, we need to understand a little of what this complete theory of quantum mechanics, developed in 1926 and the immediately following years, is all about. The package, an understanding (if that is the right word) of the quantum world based on waves, particles, chance and uncertainty, was largely developed by Bohr (taking on board ideas about probability from Max Born) and first presented to an initially bemused and sceptical scientific community late in 1927. This is the 'Copenhagen interpretation' that we have already mentioned.

In classical physics – the physics of Newton, and of all physicists before Einstein – we expect that a system of interacting particles, left to its own devices, will carry on interacting in a precise, predictable way: literally, like clockwork. But in quantum physics, the particles dissolve into a haze of uncertainty as soon as we stop looking at them. If you measure the position of an electron, for example, that doesn't mean that the electron stays there when you stop measuring its position, nor even that it proceeds along a precisely determined trajectory in obeyance to Newton's laws, so that you could calculate exactly where to look in order to be certain of finding the electron at any chosen later time. Instead, a 'wave of probability' spreads out from the position of the electron. Just after you measure the position of the electron, there is a high probability that it is still in the same place (or at the appropriate place along the relevant Newtonian trajectory), and a small, but increasing, chance that it could be somewhere else entirely. Wait for a long time before repeating the measurement, and the probability that the electron will turn up somewhere else entirely is much higher.

When we choose to measure the position of a particle accurately, said Bohr, we force it to develop more uncertainty in its momentum (where it is going); when we choose an experiment to measure wave properties, we eliminate particle features and make the position uncertain. And we can never construct an experiment that will measure wave properties and particle properties for the same quantum entity (such as an individual electron or photon) simultaneously.

Even within an atom, there are neither particles in orbit around the nucleus nor waves rippling their way around it. Instead, the right way to think of each individual electron is as a cloud of probability, thickest where the electron is more likely to be, thinner where there is little chance of finding the electron. Any individual experiment to measure the position of the electron will find it somewhere in the cloud, but that location is determined by the laws of chance, not by Newtonian clockwork. And it is *only* when the electron is measured that it 'decides' where it is; at that instant the cloud collapses to a single point, but as soon as the measurement has been made it starts spreading out again. This is the famous 'collapse of the wave function', which among other things describes (we hesitate to use the word 'explains') how, in those Japanese experiments mentioned in Chapter Ten, the 'wave' corresponding to a single electron can pass through both slits of a Young's interferometer at once, and then collapse into a single point when it arrives at the detector screen.

All this is enough to make your head spin. If it is any comfort, Einstein and Schrödinger felt the same way about it. It was in response to the Copenhagen interpretation, and in order to emphasise the absurdity, as he saw it, of the way physics was going, that Schrödinger came up with his famous 'cat experiment', which still highlights as graphically as anything can how far removed quantum physics is from the everyday, Newtonian clockwork physics that we are used to.

Schrödinger's cat

Imagine a live cat, locked in a box with some radioactive material, a geiger counter and a phial of poison. In this hypothetical experiment (a 'thought experiment'), the apparatus is set up to break the phial of poison and kill the cat if the radioactive material decays. In line with quantum rules, there will be a precisely defined time when there is an exact 50:50 chance that the radioactive material has decayed. We can open the lid of the box to find out the cat's fate. But what is the situation inside the box just *before* we take a look?

According to the Copenhagen interpretation, the quantum world does not make up its mind what state it is in until it is measured. Instead, it exists as a so-called superposition of states. The cloud of probability representing all the possible locations for a particular electron around an atomic nucleus is just such a superposition of states. In this more simple case, there are just two possible states – dead cat and live cat.[4] Quantum theory – the same quantum theory that describes how computers, the hydrogen bomb and molecular genetics work – says that just before we open the box the cat is neither dead nor alive (or, alternatively, *both* dead *and* alive), in a superposition of states.

Arguments about the problem of the cat in the box have gone on for more than half a century. One point of view is that there is no problem, since the cat itself is quite capable of deciding whether it is alive or dead, and therefore collapsing the wave function. But then, where do you draw the line? Could an ant be sufficiently self-aware to collapse the wave function? Or a computer? Or even the geiger counter that measures the radioactive decay in the first place?

Another resolution of the puzzle, seriously proposed by some physicists, is that every time the universe is faced with such a choice at the quantum level, it splits itself into two copies (or as many as there are choices). In this case, there is one universe in which you open the box to find a dead cat, and another, equally real, in which you open the box to find a live cat. With impressive understatement, this is known as the 'many worlds' version of quantum theory.

Schrödinger was disgusted with the absurdity of quantum physics highlighted by the cat in the box 'experiment'; so was Einstein. Niels Bohr, the pragmatist, didn't worry about the cat in the box as long as the theory could explain actual measurements on real systems. The absurdity has never been resolved, but quantum physics still works.

Does God play dice?

One reason why quantum physics is so securely founded as a working theory today is, ironically, the result of Einstein's opposition to it in the years following 1926. It was, indeed, in that year that he wrote, in a letter to Max Born: 'Quantum mechanics is certainly imposing. But an inner voice tells me that it is not yet the real thing. The theory says a lot, but does not really bring us any closer to the secret of the "old one". I, at any rate, am convinced that *He* is not playing at dice.'[5]

Einstein thought deeply about the implications of the quantum theory, and used to try to find thought experiments, similar in concept to the cat-in-the-box puzzle, which would demonstrate a flaw in the structure of the Copenhagen interpretation. As chief architect of the Copenhagen interpretation, Bohr would try to counter Einstein's arguments, and to show that even under the circumstances dreamed up by Einstein's fertile brain, the strict quantum rules would still apply. Because Bohr was indeed able to refute every single argument that Einstein came up with, the Copenhagen interpretation was put on an increasingly secure footing as the years went by. The debate between Bohr and Einstein lasted from 1927 until Einstein's death in 1955 – longer than the time from 1905 to 1926 during which all of Einstein's great original contributions to science were made. We'll give you the flavour of that debate with just one detailed example, the puzzle of the clock in the box.

The clock in the box

Einstein asked Bohr to imagine a box which has a hole in one wall, covered by a shutter that can be opened and closed by a mechanism connected to a clock inside the box. Apart from the clock and this mechanism, there is nothing in the box except radiation – a photon gas. The apparatus is set up (in the imagination; remember this is another thought experiment) so that at some precise, predetermined time the shutter will open and let just one photon escape before it closes again. If you weigh the box, wait for the photon to escape and then weigh the box again, then, because mass and energy are equivalent, you know, from the difference in the two weights, exactly how much energy the photon carried off with it. (Also, from the equation $E = hf$, and remembering that Einstein proved that each photon has a pure frequency, you know the frequency [wavelength] of the photon, although this doesn't enter in to this particular debate.)

According to the quantum theory, just as momentum and position are 'complementary' properties, so that it is impossible to know position and momentum precisely and simultaneously for the same particle, so are energy and time. It is impossible, according to quantum theory, to measure the precise energy of a photon (or anything else) at a precise moment in time. All you can say is that at a certain time the energy is roughly a certain amount, with the leeway of that 'roughly' determined by Heisenberg's uncertainty rule and Planck's constant. Or you can say that the energy was precisely a certain amount, at about a certain time. But Einstein's clock in the box trick seemed to allow you to say that at the precise moment the shutter opened the photon had a precise amount of energy.

Bohr showed that Einstein was wrong. He did so by looking at the practicalities of such an experiment, but still without actually carrying out the experiment. If you want to weigh the box, said Bohr, it must be connected to the outside world, perhaps by a spring balance which suspends the box in a gravitational field. Now, the rate at which the clock runs depends on its position in the gravitational field, a fact that Einstein himself had established with his general theory of relativity. But when the photon escapes, the box holding the clock moves – both because of the (tiny but not zero) recoil as the photon escapes, and because the box is now lighter, and weighs less, so the spring it is suspended from contracts slightly. The uncertainty in the rate at which the clock is running, caused by its shift within the gravitational field, restores the quantum uncertainty in line with Heisenberg's principle.

Bohr and Einstein debated the clock-in-the-box puzzle during the early years of the 1930s, and although there are other ways to refute Einstein's argument (always restoring quantum uncertainty in line with Heisenberg), Bohr was particularly delighted at being able to pull the rug from under one of Einstein's arguments against the Copenhagen interpretation by using Einstein's own most famous piece of work, the general theory. But there is still a sense in which all this debate, right up to 1955, is unsatisfactory. To an Einstein or a Bohr, it may be sufficient to work everything out in terms of thought experiments, confident in the intellectual ability to resolve the subtleties by debate. For the benefit of lesser mortals, it would be nice to have a real experiment, actually carried out in a laboratory here on Earth, to resolve the puzzle once and for all, and to decide whether Bohr or Einstein was right. Neither of them dreamed that such an experiment might be possible, and neither lived to see it done. Astonishingly, though, within thirty years of Einstein's death, and only twenty years after Bohr died, this definitive test of quantum theory was carried out in a laboratory in Paris. The answer was unequivocal; Einstein was wrong.

Einstein was wrong

Many people, when they first encounter the bizarre concepts of quantum physics such as uncertainty, and the role of the observer in determining the outcome of an experiment, instinctively react with the assumption that this must all be, in some sense, an illusion. Common sense tells us that behind the quantum scenes there is surely a clockwork of Newtonian determinism, a world of certainties and objective realities which creates the appearance of uncertainty and subjectivity to our imperfect senses and inadequate measuring apparatus.

But more than thirty years after the rules of quantum mechanics were first worked out, the physicist John Bell, working at CERN in Geneva, devised a test which could, in principle, distinguish the influence of the underlying clockwork at work – if it really was there. After a further two decades, in the 1980s, experimenters in Paris, headed by Alain Aspect, were able to carry out an actual experiment along the lines proposed by Bell. The experiments showed that common sense – and Einstein – was wrong. There is no underlying clockwork, and the strangeness of the quantum world really does have to be taken at face value.

Hardly surprisingly, this result has been described, by the Nobel laureate Brian Josephson, of Cambridge University, as the most important recent advance in physics. After Bell's death in Geneva, on 1 October 1990 at the age of only sixty-two, colleagues referred to him as the only physicist of his generation to rank with the pioneers of quantum physics, such as Max Born and Niels Bohr, for the depth of his philosophical understanding of the implications of the theory. Had he lived a little longer, he would surely have been noticed by the Nobel committee, although a rigorous understanding of his most famous (by no means his only) contribution to science does involve some rather hairy mathematics. Fortunately, though, the essence of Bell's insight into quantum reality can be gleaned without the maths, starting out from a puzzle propounded by Einstein and his colleagues in the 1930s.

Together with Boris Podolsky and Nathan Rosen, Einstein tried to show that taking quantum mechanics at face value led to a logical contradiction. Their paper, published in 1935, described what became known as the 'EPR (Einstein–Podolsky–Rosen) paradox', although it does not really describe a paradox at all.[6] The starting point for the argument is the fundamental property of the quantum world that you cannot, even in principle, measure precisely *both* the momentum *and* the position of a particle at the same time – the basis of quantum uncertainty. This is often explained in physical terms, that the act of measuring the position of a particle disturbs its momentum, and vice versa. But we cannot stress too highly, or too often, the fact that this physical explanation obscures the fact that uncertainty is intrinsic to the quantum equations – in quantum mechanics, a particle does not possess both a precise position and a precise momentum.

Einstein and his colleagues pointed out, however, that if two particles A and B interacted with one another and then flew apart, an experimenter could measure the combined momentum of the pair at the time of the interaction, and later the momentum of particle A and the position of particle B. It is then simple to calculate the momentum of particle B, whose position we already know, from the measurement made on particle A.

At least, it would be simple, if quantum particles behaved like snooker balls. The only way to save quantum uncertainty in the context of the EPR thought experiment is to allow what Einstein referred to as a 'spooky action at a distance', a communication which operates instantaneously between the two particles even when they are far apart, so that particle A can be disturbed by the measurement made on particle B – a communication which Einstein, the originator of relativity theory, could never accept, since it seemed to travel faster than light.

But the snag with all this is that it is indeed only a thought experiment, a logical argument designed to highlight certain implications of the theory. To Einstein, it was 'obvious' that there could be no action at a distance, and therefore that quantum uncertainty was an illusion. To quantum theorists, it was 'obvious' that uncertainty ruled, and therefore action at a distance had to be accepted, with the implication that once any pair (or greater number) of particles had interacted, they were ever afterwards in some sense part of a single quantum system.

None of this worried everyday physicists and engineers, who didn't care about the philosophical arguments but were happy to use the rules of quantum physics to aid their investigations of the structure of atoms and molecules, the design of lasers and microchips, and so on. But a few theorists continued to worry at the problem, and in 1952 David Bohm, of Birkbeck College in London, came up with a variation of the EPR 'experiment' that gives us some insight into what the later work of Bell and Aspect was all about.

This version of the 'paradox' deals with spin. The quantum property known as spin is not *quite* the same as the spin we are familiar with – the spin of a top, or the spin of the Earth in space. An electron, for example, has to 'rotate' *twice* in order to return to the state in which it started. But such subtleties need not worry us here. What matters is that particles do have spin, and that, like its classical counterpart, the amount of spin (the 'angular momentum') associated with the particles in an interacting system is conserved.

So, if a hypothetical particle that has zero spin (no net angular momentum) decays into two particles that do have spin – an electron and a positron, perhaps – then the spins of these particles must be equal and opposite. That would be all well and good if the spin of a particle were an intrinsic property, like the mass of a billiard ball – part of the basic clockwork underpinning quantum physics. But it is a fundamental feature of quantum mechanics that quantum entities obey *probabilities*, not hard and fast certainties.

In the case of an individual electron, this means that the electron can have either spin 'up' or spin 'down' (positive or negative spin) with equal probability. If nobody measures the spin, the electron is said to exist in a superposition of states, a mixture of both possibilities. But when a measurement is made there is a collapse of the wave function and the experimenter gets a definite answer, either up or down. Repeat the experiment a large number of times, and half the time the measurement will come out spin up, the other half it will come out spin down.

This is where the plot begins to thicken. Our two particles, flying apart from the decay of a spin-zero particle, do not actually have well-defined spin, according to quantum theory, until it is measured. They each exist in the mixed superposition of states. We could wait, in principle, until they were light years apart, and then measure the spin of one of the particles. At that instant, the wave function would collapse, and for the first time the particle could be said to have a definite spin. By the conservation of angular momentum, the wave function of the other particle, light years away, must collapse *at the same instant* into the state with the opposite spin.

Bohm's version of the EPR paradox makes the situation even clearer, but it is still only a thought experiment. The conceptual breakthrough that led to a practical experiment to measure these bizarre effects came with the publication of a paper by John Bell in 1964. The Bell test is even more complicated than the picture we have just sketched, because in fact the spin of a particle such as an electron or a proton can be measured independently in three directions at right angles to one another – 'orthogonal', in the jargon of the trade. But this apparent complication is actually what makes a practicable test possible. You can measure the spin of particle A along the X-axis (in some arbitrarily chosen coordinate system), and if quantum theory is correct, this measurement also affects the spin of particle B in the X direction, but not in the Y or Z directions. And you can measure the spin of particle B in the Y direction, say, without affecting the X and Z measurements for particle A. In this way, it is possible to glean information about complementary spin states in both particles.

What Bell showed in 1964 was that measurements of these orthogonal spins on large numbers of real particles could in principle distinguish between underlying clockwork and action at a distance. If Einstein was right, and the particles really did have an intrinsic spin all the time, then in a Bohm-type experiment the number of pairs of particles in which both are measured to have positive spin in both the X and Y directions (XY positive) is always less than the combined total of measurements showing XZ and YZ measurements *all* to have positive spin (XZ positive + YZ positive). This is 'Bell's inequality'.

If quantum theory is right, and the spin is undetermined until a measurement is made, the inequality would work the other way round (that is, with XY positive greater than XZ positive + YZ positive). Because the way Bell set up the argument was with the assumption of underlying clockwork and objective reality (the way Einstein would have liked it), if Bell's inequality is *violated*, then quantum theory is correct and we have to live with action at a distance.

This is a very powerful piece of reasoning, which Bell expressed in a very elegant mathematical way. In an interview with Paul Davies on BBC radio in 1985, Bell said that he got the basic equation into his head and out on to paper within a single weekend – but he went on to say that he had been thinking intensely around all these questions for several weeks at that time, and that the EPR paradox had been one of the puzzles at the back of his mind for years before that.[7]

It certainly took more than a weekend for Bell's conceptual breakthrough, translating the EPR thought experiment into something that could actually be carried out in principle, to become a practicable working experiment. The Aspect experiment was the definitive version of several attempts to measure the Bell inequality. It actually involves measurements of the polarisation of pairs of photons, rather than the spins of material particles, but the principle is the same and it does provide a direct measurement of the Bell inequality. In a series of runs of the experiment in the early 1980s, Aspect's team established that Bell's inequality is violated – that is, quantum theory, including the spooky action at a distance that Einstein hated so much, is correct.

The implications have yet to be absorbed by theoretical physicists and philosophers, let alone by the rest of the world. But they are profound indeed. The concept of an influence travelling faster than light to link particles in a cosmic web goes deeper than it may seem from the simplified picture of a single particle decaying into a positron and an electron, since according to big-bang theory (itself rooted, remember, in the general theory of relativity) *all* particles originate from a common set of interactions at the birth of the universe. There is no prospect, however, of using the cosmic network to send signals faster than light, since measurements of particle A produce random results in line with the quantum rules of probability, and therefore cause random changes in the state of particle B.

The random changes in B may be different from the random state it would have if we left A alone, but, as Heinz Pagels has expressed it, 'if we asked out of all the things in this universe which one, if altered in a random way, would remain unchanged, the answer is: a random sequence. A random sequence changed in a random way remains random – a mess remains a mess.'[8]

For those still worried by the implications, though, Bell had a mischievous response. He was a quiet, gentle man with a delightful sense of humour, liked as well as admired by his colleagues, but with his own ideas about what really mattered in life. There is a story that he once accepted an invitation to attend a physics meeting in Denmark, rather than a more prestigious gathering taking place simultaneously in another country, because he knew there was a shop in Copenhagen where he could get foot-shaped shoes. Such a man was hardly likely to be afraid of airing unconventional views, or to worry if he strayed outside the bounds of received wisdom.

In that same BBC interview, he did both. Bell explained that there is still a way to escape the inference of action at a distance in the light of the outcome of the Aspect experiment. But it involves absolute determinism in the universe – the complete absence of free will.

According to the interpretation of events we have outlined here, the experimenter is free to decide what is measured and when. The measurement then influences events at a distance, in the way that Einstein abhorred. But just suppose that the world is super-deterministic, with not just inanimate nature running on behind-the-scenes clockwork, but with our behaviour, including our belief that we are free to choose to do one experiment rather than another, absolutely predetermined. If *everything* is predetermined, including the 'decision' by the experimenter to carry out one set of measurements rather than another, the difficulty disappears. There is no need for a faster than light signal to tell particle A what measurement has been carried out on particle B, because the universe, including particle A, already 'knows' what that measurement, and its outcome, will be.

And that, perhaps, is the best measure of the way in which the Aspect experiment and Bell's inequality change our view of the universe. The *only* alternative to quantum probabilities, superpositions of states, collapse of the wave function and action at a distance is that absolutely everything is predetermined. 'For me,' said Bell, 'it's a dilemma. I think it's a deep dilemma, and the resolution of it will not be trivial; it will require a substantial change in the way we look at things.' And, unlike those physicists who now believe that we are on the brink of finding a 'theory of everything' that will wrap all of physics up in one neat package, Bell was prepared to say only that he thought there would be theories that are better than the ones we have today, and that even 'quantum theory is only a temporary expedient'.

Einstein, like Bell, died believing that quantum theory is only a temporary expedient, and that a better description of the universe must one day be found. He may yet be proved right. But until we get that better theory, quantum mechanics remains the best description of reality that we have, and the paradoxes of cats and clocks in boxes, as well as the spooky action at a distance, are the price we have to pay for a working understanding of phenomena as diverse as lasers, nuclear power, computer chips, virtually all the modern understanding of chemistry (which depends on understanding the behaviour of electrons in the outer parts of atoms) and, through biochemistry, our understanding of genetics, DNA and life itself.

It is often suggested that the last thirty years of Einstein's life were in some sense wasted, in spite of his earlier achievements, because his opposition to the quantum theory cut him off from the mainstream of scientific development between 1925 and 1935. But this misses the point that simply by providing such carefully reasoned opposition to the Copenhagen interpretation and forcing Bohr, Bohm and others to come up with new ideas in response to his arguments, Einstein ensured that the theory was put on a more secure footing, more quickly, than would otherwise have been the case. That was certainly not his intention; but in the role of quantum opponent he made yet another invaluable contribution to science. By the end of the 1930s, however, developments in the world outside the scientific community were overshadowing even this fundamental debate about the nature of reality.

Chapter Thirteen

The Final Years

A few days before leaving Europe for the last time, Einstein received a telegram from Abraham Flexner which read: 'Have been authoritatively requested to ask you to enter this country quietly and inconspicuously and disembark at Quarantine. To make no statement, give no interview on any subject to newspapers. Please keep cable absolutely confidential. Warmest greetings. Flexner.'[1]

Thus began the last phase of Albert Einstein's life; a peaceful existence in the refined atmosphere of Princeton, though he would be kept on a tight rein by authorities of one variety or another.

Princeton in 1933 was not so different from the Princeton of today. It is a sleepy little town no more than an hour's drive from Manhattan. It has one main street lined with picturesque wooden houses and neatly trimmed gardens. Totally dominated by the university, it could perhaps be a less grand version of Cambridge, although the town's buildings are from the nineteenth century rather than the Middle Ages and the whole atmosphere is more isolated and more laid-back.

When Einstein arrived, the Institute of Advanced Study was still located on the site of the university. It did not have its own buildings until 1938, when the whole establishment moved to the edge of town and the splendid buildings which house it to this day.

Although Einstein always claimed that he would return to Oxford and perhaps other European cities as soon as things had settled down, it is quite likely that, even upon his arrival in

America, he realised that this would be his final home. During the course of the next two decades, Einstein only left America once, visiting Bermuda in 1935 in order to make an official application for permanent residency in the United States. For the vast majority of the time, he lived in relative isolation from the rest of the world, cocooned within the cosy environment of his New Jersey home.

Princeton perfectly suited Einstein during his later years. The institute was geared towards pure research. In company with the other professors there, he was free from all teaching commitments because there were no students, save a tiny number of postdoctoral research students. He had the best facilities at his disposal, the peace and solitude of the area, a comfortable home, no financial worries and a temperate climate.

However, the locals took some time to get used to the influx of European scientists on their doorstep. Many of the older residents had a definite streak of anti-Semitism and there was more than the occasional raised voice concerned that so many foreigners in their midst would in some way disrupt the established way of life in the town. Later on Albert Einstein came to be seen as a feature of the area, and by 1992, everyone in Princeton seemed to have a story to tell about 'old Albert'. Everybody had a grandfather or an ancient uncle who had known Einstein. An old cab driver claimed to have been employed to follow Einstein to and from work each day in case the elderly scientist decided that he had had enough of walking to and from the institute, while deep in thought. There was the old woman who owned the hardware store in the centre of town where Mrs Einstein used to shop and there were still a few, but rapidly diminishing number of eminent colleagues of the great man, still at work at the Institute of Advanced Study.

There are numerous stories of Einstein's early days at Princeton, stories which have become legends, told and retold to interested visitors. One recounts that a little girl who lived along the road from the Einsteins' house was having trouble with her maths homework. Hearing that a great mathematician lived nearby, she sought him out. Growing concerned by the daily disappearance of her daughter, the child's mother one afternoon followed her to a house at 112 Mercer Street, only to find to her embarrassment that her daughter had asked the most famous scientist in the world to help her with her maths problems.

Another favourite anecdote illustrates Einstein's growing absent-mindedness. Catching the bus into town one day, he seems to have taken his time working out the correct change to give the bus driver. After fumbling around with a small collection of dimes and quarters, the impatient driver took the money out of Einstein's hand and sorted it out for him. 'Not much good at sums, then?' he said as he handed Einstein his ticket.

On another occasion, the secretary of the dean's office at the institute received a telephone call from someone asking to speak to the dean. When the secretary informed the caller that the dean was out, the voice on the end of the line said: 'Well, perhaps you can help me. Could you please tell me where Dr Einstein lives?' The secretary had been specifically instructed not to give out Einstein's address and telephone number, and told the caller so.

There was a silence at the end of the line, then the voice returned, sounding shaky. 'Oh dear!' the voice said. 'This is terribly embarrassing. This is Professor Einstein. I've forgotten where I live.'

It is difficult to overestimate the level of celebrity bestowed upon Einstein throughout the second half of his life. When he arrived in America he was already seen as an icon of the twentieth century, not just an eminent scholar, but a great figure of world history. Flexner knew what a catch Einstein had been and realised that the evolution of the institute had largely depended on this. Consequently he was overprotective of Einstein, an attitude that caused a number of disputes between the two men.

The most serious clash came in January 1934, when Einstein was invited to visit President Franklin Roosevelt at the White House. The invitation was sent to the institute and was intercepted by Flexner. Without even informing Einstein, Flexner took it upon himself to reply that, regrettably, Professor Einstein would be unable to make the visit. Einstein had heard through a friend that the president was about to invite him to Washington and had expressed eagerness to meet Roosevelt. When, some time later, this same friend heard that Einstein had declined the invitation, he was surprised enough to get in touch with Einstein. Thus, Flexner's deception was uncovered.

Flexner may have feared that the president would entice Einstein away from the institute. The Einsteins finally dined at the White House on 24 January and stayed the night. There is no record of the evening's conversations, but Einstein described the meeting as convivial and was impressed with Roosevelt's command of German. The president merely wanted to meet the world's greatest scientist, who had chosen to reside in America. He also wanted to learn as much as he could about the situation in Europe as Einstein saw it and to get his views on German-US relations.

The Einsteins had been in the United States for less than a year when their peace and happiness were shattered by the first of a series of shocks. In the summer of 1934, they heard from France that Einstein's stepdaughter Ilse was gravely ill.

Ilse was suffering from cancer and had stipulated that her mother should not be informed until it was clear that her condition was critical. Upon arriving in Paris, where Ilse was being nursed by her sister Margot, Elsa was horrified at the sight of her daughter wasted away by the disease.

Ilse's illness had been diagnosed as incurable, a wrong diagnosis according to one of Einstein's early biographers, Antonina Vallentin. Elsa sent for a German doctor to be flown to Paris and a stream of French doctors came and went. All to no avail, Ilse died soon after Elsa's arrival, and the distraught mother returned to America. Margot and her husband soon followed and settled in Princeton. Albert and Elsa decided to leave Princeton for a spell and rented a ranch in Connecticut.

It seems to have been an idyllic break for Elsa. The rented house was beautiful and came with over 20 acres of land, a swimming pool and a tennis court. Einstein was able to work in the peace and solitude of these surroundings and Elsa tried to overcome the pain of her loss. The countryside of Connecticut was very different from the woods around their previous home in Caputh. Instead of the dense woodland and gurgling brooks, the lake and the tiny village, the Einsteins found themselves amid the fields and early autumn woods of the northeast, but it was quiet and unmolested by the growing noise and dirt of American city life.

Elsa was also unwell. The first serious symptom appeared in the summer of 1935, as the family were moving from a rented house near the university to their own property at 112 Mercer Street. During the chaotic weeks of the move, Elsa noticed that her eye had become swollen. At first she hoped that it was a passing inflammation. When after a week the swelling had not gone down, she visited a consultant oculist in New York. She learned that the eye swelling was a symptom of a serious disorder of the heart and kidneys.

Elsa took the decision not to undergo surgery or to move to hospital. She wanted to stay in the new family home at Mercer Street with her beloved husband. As a compromise measure she was given a course of medication and forced to stay bedridden and completely immobilised.

During this period Helen Dukas was an immeasurable help. Throughout Elsa's time in Europe visiting her dying daughter, Miss Dukas had increasingly taken on the role of housekeeper for Einstein. Now, with the mistress of the house confined to bed, the responsibility of looking after Einstein's domestic needs as well as secretarial work fell permanently on her shoulders. It forged ties between Einstein and his secretary and paved the way for their close working relationship after Elsa's death. Helen Dukas remained Einstein's closest helper until the day he died and she played a significant role in Einstein's literary affairs thereafter.

Meanwhile, the drastic attempt at a cure for Elsa appeared, for a while, to work. By the spring of 1936, she seems to have made the beginnings of a recovery and the couple decided to take a summer trip to Saranac Lake, about 300 miles north of Princeton, where she could convalesce. However, Elsa was not happy there. In letters to a family friend, Leon Watters, she complained that her husband could spare little time for her. Einstein continued to be absorbed in his work even when the couple were alone.

It is conceivable that Einstein did not fully realise the seriousness of his wife's illness. It is certainly true that she did her utmost to keep the details from him and probably encouraged the view that she would soon make a full and complete recovery. However, for all his absent-mindedness in his later years, Einstein was certainly no fool when it came to domestic situations. After all, he had been through a number of illnesses himself and must have known that Elsa was very sick.

There is also the curious fact that Elsa contradicts her unhappiness in other letters, this time to one of her closest, friends, Antonina Vallentin. At several points Elsa declares that Einstein is full of sadness at her illness and that she did not realise that Albert loved her so much.

After a brief spell of better health during that summer, Elsa's condition rapidly deteriorated upon her return to Princeton. She died in December 1936.

Einstein and Elsa had lived together for the best part of two decades and the couple had been through much together. The good times in the early Berlin days and Einstein's sudden international fame, the adventures together in the Far East, America and England and the bad times, the loss of their beautiful little house in Germany, the years in exile and the death of Ilse. Now, Einstein had lost a second wife, the first through incompatibility and divorce, the second through death.

According to a number of people, Einstein faced his loss with a lack of sentiment veering on coldness. Soon after Elsa's death he wrote to his friend and colleague Max Born: 'I have settled down splendidly here. I hibernate like a bear in its cave, and really feel more at home than ever before in all my varied existence. This bearishness has been accentuated further by the death of my mate, who was more attached to human beings than I.'[2]

This letter is frequently quoted to support the assertion that Einstein did not fully value his wife. He was undoubtedly unaware of all that Elsa had done for him, but he was a loner. He enjoyed the company of other intelligent human beings but he carried a reserve of independence within him throughout his life. According to others close to Einstein, the loss of Elsa affected him more than he could express.

It could be argued that Einstein was incapable of love in the sense that most people believe they understand the word. For all his artistic and poetic inclinations, when it came to deep feelings, Einstein had a definite blind spot. He could feel for the human race as an entity but for the individual only with reservation. He saw art and the expression of inner emotions as a creative thing, the fulfilment of a deep-rooted urge, not as an out pouring of deeply buried and intangible emotions such as love. It could even be said that Einstein could visualise hate more readily than he could understand love.

Whether to suppress his pain over Elsa or simply out of habit, from 1936, Einstein redoubled his efforts to find a path through the intricacies of relativity and quantum mechanics in a search for unification. As Europe veered ever closer to war, he deliberately submerged himself in a world where few could follow. Einstein had many visitors in Princeton, both friends and colleagues, as well as a seemingly endless stream of people who all wanted his help with one cause or another. Postal requests to deliver talks for charitable organisations arrived at 112 Mercer Street almost daily. And throughout the prewar years refugees kept arriving in the United States, homeless scientists and other intellectuals creating a brain drain from mainland Europe. Einstein played a major role in helping them find places at American universities and research establishments. Several of these immigrants would, in only a few years, play a key role in the Allies' atomic-bomb programme. In the years leading up to the outbreak of war, Germany lost six Nobel Prize winners, and such great names in physics as Enrico Fermi, Leo Szilard, Otto Stern and Lise Meitner moved westwards to aid Germany's eventual defeat.

As Einstein delved further into the obscure world of the unified theory, across the Atlantic breakthroughs were being made in the field of atomic physics. In little more than a decade these discoveries were to lead to the atomic bomb and the beginning of the Atomic Age.

The new era really began in 1933, when Leo Szilard had the inspired idea that a chain reaction could be initiated by the bombardment of an element with neutrons, if each bombarded atom in that element then emitted two neutrons. However, it was not until 1938 that several European scientists working independently stumbled upon nuclear fission. The Italian Enrico Fermi seems to have been the first. But Madame Curie's son-in-law, Frédéric Joliot, also observed the phenomenon, as did Otto Hahn and Fritz Strassman in Germany.

Having completed their experiments, Hahn and Strassman sent their findings to Hahn's former colleague Lise Meitner, who had left Germany to live in Holland. She was visiting Stockholm for Christmas and received Hahn and Strassman's findings there. During a walk in the woods near the house of her nephew, a fellow physicist with whom she was staying, Meitner discussed with him the ramifications of Hahn's findings and formulated the theoretical background to the German physicist's experiments.

Niels Bohr brought the news to America that nuclear fission had occurred and that its theoretical explanation had been derived. He made the announcement at the Fifth Washington Conference on Theoretical Physics in 1939. From this forum the news spread like wildfire through the scientific community.

For many historians it has been a source of amazement that the Nazis did not catch on to the military uses of atomic energy. Half the world's scientists were talking about the breakthrough in the field of nuclear physics and discussing its uses. How was it that the German government, usually quick to realise the military potential of any new development, were so slow on the uptake?

Firstly, Hitler seems to have been totally uninterested in the whole phenomenon of atomic energy. Secondly, many of Germany's leading scientists deliberately suppressed news of their discoveries. They were loath to let the Nazis in on the secret for fear of the consequences. Thanks to the efforts of Heisenberg, in particular, the Nazis fell behind the Allies in the development of atomic weapons.

It is important to remember that the Allies did not know that they had this advantage until close to the end of the war. It was their fear that the Nazis were ahead of them that motivated the setting-up of the Manhattan Project and the channelling of huge resources into the development of atomic weapons in the United States from 1941.

The role Einstein played in the establishment of the Allies' atomic-weapons programme has gone down in folklore, but the facts behind the legend have often been blurred and the part he played in the scheme of things has been totally misinterpreted.

The orchestrator of the campaign to establish the Allies' atomic-weapons programme was the Hungarian physicist Leo Szilard. He wanted to alert President Roosevelt to the need to beat the Germans to the bomb. But Szilard realised that it would take the influence of an Einstein to make any impression on the head of state. Szilard was way ahead of his time – he was worried about the potential of a German atomic-weapons programme as early as 1938, before the war had broken out in Europe and the best part of four years before America had become involved in the fighting. He knew also that the world's main source of uranium, the essential element needed in any atomic research, was to be found in the Belgian Congo. If Germany were to invade Belgium, the Nazis would have exclusive use of this uranium, preventing future enemy states from developing their own programmes. Szilard decided in July 1939 that he should visit Einstein.

During that summer Einstein was staying with friends on Long Island. Szilard and a sympathetic colleague, Eugene Wigner, discovered the address and turned up out of the blue to explain their fears to Einstein.

Einstein was staggered by the idea of utilising nuclear fission in an atomic bomb. This reaction has surprised many commentators. How could a man of Einstein's vision and unparalleled ability have failed to realise the military potential of the nuclear chain-reaction process, which had, after all, been suggested by Szilard as early as 1933?

The most likely explanation for this is simply that Einstein was more isolated from the world than many realised. Of course he heard news of the latest work within the scientific community and subscribed to the scientific journals of the time, but may not have read much outside his immediate concerns. His own work during the 1930s and 1940s had very little to do with mainstream physics and for the last twenty years of his life he was more or less working alone on the unified theory. The chances are that he had not followed the arguments about the uses of nuclear fission.

Soon after the end of World War II, Einstein said of the news Szilard had brought him: 'I did not, in fact, foresee that it would be released in my time. I only believed that it was theoretically possible.'[3]

Einstein was not alone in his scepticism over the application of nuclear fission. The physicist Frederick Lindemann, who had been Einstein's host on many of his frequent visits to England, is said to have expressed astonishment that the universe could be constructed in such a way as to allow such power to fall into our irresponsible hands. He was once described by a colleague as regarding most of humankind as furry little animals.

As is well known, Einstein agreed to act as the spokesman for the scientific community and to sign a letter drafted by Szilard aimed at persuading the president to take action over the potential uses of nuclear fission. The finished letter to Roosevelt, dated 2 August 1939, read:

> Sir:
>
> Some recent work by E. Fermi and L. Szilard, which has been communicated to me in manuscript, leads me to expect that the element uranium may be turned into a new and important source of energy in the immediate future. Certain aspects of the situation seem to call for watchfulness and, if necessary, quick action on the part of the administration. I believe, therefore, that it is my duty to bring to your attention the following facts and recommendations.
>
> In the course of the last four months it has been made probable – through the work of Joliot in France as well as Fermi and Szilard in America – that it may become possible to set up nuclear chain reactions in a large mass of uranium, by which vast amounts of power and large quantities of new radium-like elements would be generated. Now it appears almost certain that this could be achieved in the immediate future.
>
> This new phenomenon would also lead to the construction of bombs, and it is conceivable – though much less certain – that extremely powerful bombs of a new

type may thus be constructed. A single bomb of this type, carried by boat or exploded in a port might very well destroy the whole port together with some of the surrounding territory. However, such bombs might very well prove to be too heavy for transportation by air.

The United States has only very poor ores of uranium in moderate quantities. There is some good ore in Canada and the former Czechoslovakia, while the most important source is in the Belgian Congo.

In view of this situation you may think it desirable to have some permanent contact maintained between the administration and the group of physicists working on the chain reaction in America. One possible way of achieving this might be for you to entrust with this task a person who has your confidence and who could perhaps serve in an unofficial capacity. His task might comprise the following:

a) To approach government departments, keep them informed of further developments, and put forward recommendations for government action, giving particular attention to the problem of securing a supply of uranium ore for the United States.

b) To speed up the experimental work which is at present being carried on within the limits of the budgets of the university laboratories, by providing funds, if such funds be required, through his contacts with private persons who are willing to make contributions for this cause, and perhaps also by obtaining the cooperation of industrial laboratories which have the necessary equipment.

I understand that Germany has actually stopped the sale of uranium from Czechoslovakian mines which she has taken over. That she should have taken such early action might perhaps be understood on the ground that the son of the German Undersecretary of State, von Weizsäcker, is attached to the Kaiser Wilhelm Institute in Berlin, where some of the American work on uranium is now being repeated.

Yours very truly,
A. Einstein.[4]

The intermediary for the letter was an economist, Alexander Sachs, who was known to have considerable influence with the president. However, Sachs needed to wait for the appropriate moment to pass the letter on to Roosevelt. In the event this was not until 11 October.

After reading Einstein's letter, Roosevelt declared: 'This requires action.' By the end of the same evening he had set up a small group to investigate the potential uses of the fission process. From that moment the road to Hiroshima lay open.

Einstein had faced a serious moral dilemma when Szilard had asked him to give his support to the creation of an atomic-weapons programme. In the space of a few years he had transformed his political view from extreme pacifist to advocate of atomic weaponry. But this change had not come about without a great deal of soul-searching. If the Allies did not create the atomic bomb, then, sooner or later, the Nazis would. Passive resistance could not work in that situation. Einstein therefore signed the letter.

Because of this, Einstein was later perceived by many as the father of the atomic bomb. This is entirely wrong. Einstein made two contributions to the production of the bomb. His first was the formulation of the equation $E = mc^2$; the second, his effort to alert the US president to the danger that the secrets of nuclear fission might fall into enemy hands. Contrary to popular myth, Einstein had absolutely nothing to do with the Manhattan Project to build the bombs dropped on Hiroshima and Nagasaki. He was not present at any nuclear test.

Nonetheless, his association with the horrors of the atomic bomb was, in later years, a source of great sadness for Einstein, and he felt that his role in the process had been completely misinterpreted and exaggerated. There is no doubt that by writing to Roosevelt, Einstein accelerated the development of the atomic bomb. But the mechanism for its production was already largely in place and the willingness of the military to make the bomb came very quickly.

The calculations which showed the feasibility of the bomb's construction had already been done in England. In fact, the British government was already seriously investigating the weapon's potential when Roosevelt heard of the process. It was soon realised that the industrial resources needed for the project would not be available in Britain on the brink of war. The project would have to be set up in the United States.

Roosevelt's investigative group was called the Briggs committee, headed by Dr Lyman Briggs, director of the US Bureau of Standards. The first meeting took place ten days after Roosevelt received Einstein's letter. It was attended by Szilard, Wigner and Fermi, who was later to be instrumental in the development of the hydrogen bomb. Albert Einstein was not present.

The committee decided at the first meeting that a research programme should be created in American universities to investigate the fission process and to develop the best method for converting the theory into a workable weapon. Einstein was formally invited to join the committee, but declined.

Thus ended Einstein's involvement with the atomic-weapons programme. After initiating the project, by a combination of his own wishes and the suspicions of the authorities, Einstein played no further role. While Oppenheimer, Fermi, Szilard and the rest grew increasingly involved with the bomb's development and eventual use, Einstein returned to isolation in Princeton, knowing little more than the man in the street about the development of the most destructive weapon the human race had ever created and the most ambitious technological effort in history.

In recent years questions have been raised of why it was that Einstein was not intimately involved in the Manhattan Project. After all, it has been said, Einstein was universally recognised as the greatest scientist alive; surely he would have been able to contribute enormously to the eventual success of the project?

The answer is complex. First, the fact is often overlooked that Einstein did not want to be involved in the development of the atomic bomb. He believed strongly that the Germans should not be allowed to develop the bomb first, but he would have found practical involvement in the project too much at odds with his political and pacifistic ideals.

Secondly, although Einstein was deeply involved with quantum mechanics throughout the 1930s, his work centred around the unification of relativity and quantum mechanics – some way removed from the practical applications of nuclear fission. It is therefore true to say that his contribution to the project would have been of limited value. And by 1942, when the Manhattan Project was established at Los Alamos, Einstein was sixty-three and far from healthy. He could not have uprooted himself from Princeton and moved with the project to Los Alamos.

Even if Einstein had been willing and able to become involved in the Manhattan Project, he would have been prevented from doing so because of the fears of the Federal Bureau of Investigation. Obsessed with searching for communist sympathisers, the FBI had been investigating his activities ever since he first took up residence in the United States. The agency had branded Einstein as an 'extreme radical' and had, by 1940, compiled a 1,500-page dossier on him. There is therefore little doubt that by the autumn of 1939, J. Edgar Hoover, the bureau's chief, would have given Roosevelt his strongest recommendation not to allow Einstein to be party to any military secret. By this time Einstein was seen as an extreme security risk.

The notion that Einstein was a communist sympathiser, in the sense meant by Hoover, is of course ridiculous. However, he was a man perceived by the government as being led along political pathways far from his own views and had always been susceptible to the manipulation of the fanatic. For example, Einstein was honorary president of an organisation called the League Against Imperialism and for National Independence, a communist organisation that, in 1931, campaigned for the release of two known communist activists. Einstein wrote letters to several US senators, including William Borah, the chairman of the Select Committee on Foreign Affairs, asking him to intervene. These were passed on to the FBI, who were certainly hunting around for something to pin on Einstein. He was, after all, a foreign scientist living in the United States at a time of international unease. He was famous for his left-of-centre politics and extreme pacifist views. As far as the FBI was concerned, he had to be watched.

Consequently, for a combination of reasons, Einstein played no role in the Manhattan Project. Scientific colleagues working at Los Alamos would visit Einstein in Princeton when they were in the area, but they were forbidden by the military to speak about their work to anyone outside the project. The military, under the command of General Groves, also separated the scientists working on the bomb into self-contained groups. Communication between the groups was kept to an absolute minimum, thereby reducing the value of any sensitive aspect of each group's work.

Despite the clandestine methods by which the FBI worked, they may well have been correct in keeping nuclear secrets from Einstein. Either through naivety or carelessness he would have been a potential security risk. He was just not worldly enough to understand the concept of censorship or secrecy. He was far too trusting of others and far too vulnerable to be entrusted with state secrets at a time of crisis.

Einstein did, however, contribute to the war effort in other ways. From June 1943, he was employed by the Bureau of Ordnance, the Special Service Contract department of the US Navy. He worked part-time for the Navy from his study in Princeton and was visited every other Friday by George Gamow, who was selected by the Navy because he was a fellow scientist who had known Einstein on a personal level before the war.

Einstein's job was simple. Presented with plans for new types of weapons and delivery systems, he would point out flaws and make suggestions on how the ideas could be improved and developed. It was a job similar to that at the Swiss Patent Office and served as light relief away from the intensity of the unified theory. It was also a job which did not require security clearance since most of the weapons were at a very early stage of development, often little more than ideas.

Einstein was paid $25 per day for his services, the maximum paid by the Navy for such contractual work in 1943.

On the morning of 6 August 1945, the first atomic bomb was dropped on the Japanese city of Hiroshima. Some 70,000 Japanese citizens died instantly and up to 100,000 perished later from burns and radiation sickness. The Allied war effort to produce a usable atomic weapon had worked.

The first Einstein knew of the event was a radio broadcast announcing the delivery of the bomb. His reported reaction was a stunned '*O Weh*!' ('Oh, horrible!')

It was only after the war that the Allies realised how far the Germans had been from developing their own atomic weapon. Heisenberg, Hahn and von Laue – who, by 1945, were all interned in Farm Hall, the postwar debriefing centre for German scientists – were stunned by the news of the Hiroshima bombing.

After the end of World War II, atomic weapons became an essential component of the military capacity of five of the world's most powerful nations. The secrets of the Manhattan project were the joint property of the US and British governments. France was then let in on the secret, and within a few years the USSR and China both managed to develop their own nuclear programmes.

Seeing this nuclear-weapons proliferation, Einstein realised that the world was heading for disaster in a nuclear inferno. From the end of the war until his death in 1955, Einstein campaigned unceasingly for the abolition of nuclear weapons. He delivered speeches whenever his health would allow him to travel. He rarely left the environs of Princeton, but made occasional brief trips to New York to give short talks on the subject so close to his heart – the dangers of nuclear-weapons proliferation.

Einstein always considered the communist threat a nonsense. He did not perceive Stalin to be another Hitler and did not view the Russian people as having the same empire-building impulse and aggression as the Germans. He saw the growing conflict of the two world superpowers as a cold war without any ethical grounding – a complete folly. There was no monster at the door against whom to defend at all costs, merely a clash of ideals, a conflict of world views, and in such a climate Einstein had no hesitation in returning to his prewar pacifism.

In this cause Einstein forged closer links with his friend in England, the philosopher and mathematician Bertrand Russell. The two of them made many plans to generate publicity for the pacifist cause and lead the way as one in their fight against what they saw as being the absurd state the world had reached after what should have been the lesson of World War II.

One of Einstein's most significant efforts in the arena of

anti-nuclear campaigning was his key role in an organisation called the Emergency Committee of Atomic Scientists. He was the president and chairman of trustees of the committee and acted as the big name to attract much-needed public interest. The brief of the organisation was simply to educate the general public about the true nature of atomic weapons in the hope that people would soon realise that their governments were behaving immorally. To this end Einstein delivered talks and was interviewed for the newsreels and radio. He wrote articles for national newspapers and contributed to magazines and leaflets produced by the committee. However, despite having a plethora of names behind it, including Linus Pauling, Leo Szilard and the German-American physicist Hans Bethe, the committee failed to make much of an impression on the awareness of the general public.

During the last ten years of his life, Einstein succeeded in finding some of the peace and isolation which he had always craved. He divided his time between his twin passions, physics and the moral dilemmas of the world.

Although he continued to pursue the Holy Grail of the unified theory, the work was really leading him into more and more isolated backwaters on the extreme edge of physics and into a realm where very few could follow. In the world of politics changes in the postwar age stepped up a pace and society continued to take the most violent and wasteful path. Einstein maintained his own equilibrium and stuck to the scientific and moral frameworks which had always served him well.

In Princeton, Einstein continued to divide his life between the institute and his own study in Mercer Street. He had officially retired in April 1945 but, for Einstein and a diminishing collection of other luminaries at the institute, special retirement arrangements had been made. Einstein kept his office at the institute and still worked there several times a week; he still received a salary and had full use of the facilities of the establishment.

It was into this peaceful environment that a flattering and surprising offer came Einstein's way, in November 1952, when he was proposed as the new president of the state of Israel.

Israel had been created in May 1948 and its first president had been the great Zionist Chaim Weizmann. When Weizmann died on 9 November 1952, there was no obvious successor. The title was really an honorary one and carried little power; the president was merely a figurehead and the country was actually run by the prime minister, who at the time was David Ben-Gurion.

Einstein's name was first put forward for the presidency by the Israeli newspaper *Maariv* in an attempt to gauge public opinion. The suggestion was immediately and warmly supported by Ben-Gurion, who all but idealised Einstein. 'If we are looking for a symbol,' Ben-Gurion said, 'why not have the most illustrious Jew in the world, and possibly the greatest man alive – Einstein?'

Einstein and those close to him had not taken the suggestion at all seriously, though Einstein had wondered, if the stories were in fact true, how he was to decline gracefully.

Before the idea was proposed to the Knesset, the Israeli ambassador to the United States, Abba Eban, was instructed to phone Mercer Street to sound out Einstein's reaction informally, and the *New York Times* got hold of the story. To the ambassador, Einstein declared that he was touched and flattered but that it would be impossible for him to accept the gracious offer. To the *New York Times* he made no comment.

The message was passed on to the Israeli government but they were not going to give up that easily. Ben-Gurion made a personal appeal to Einstein outlining the situation and making it clear that the position would in no way distract the scientist from his great work, that the public responsibilities connected with the position were minimal and that the country and the Israeli people would be greatly honoured by his acceptance.

The formal invitation duly arrived. After considerable thought, Einstein realised that he could not accept the offer, whatever Ben-Gurion said. Einstein's written response read:

> I am deeply moved by the offer from our state of Israel, and at once saddened and ashamed that I cannot accept it.

All my life I have dealt with objective matters, hence I lack both the natural aptitude and the experience to deal properly with people and to exercise official functions. For these reasons alone I should be unsuited to fulfil the duties of that high office, even if advancing age was not making increasing inroads on my strength.

I am the more distressed over these circumstances because my relationship to the Jewish people has become my strongest human bond, ever since I became fully aware of our precarious situation among the nations of the world.

Now that we have lost the man who for so many years, against such great and tragic odds, bore the heavy burden of leading us towards political independence, I hope with all my heart that a successor may be found whose experience and personality will enable him to accept the formidable and responsible task.[5]

Thus the matter was concluded. Itzhak Ben-Zvi was elected Israeli president and Einstein returned once more to his quiet life in Princeton.

By 1952, Einstein's health was rapidly deteriorating. He had been severely weakened by his illness of 1917 and had experienced recurring bouts of sickness, crippling stomach pains and extreme tiredness. In 1945, he had agreed to undergo an investigatory operation. No conclusion was reached. Three years later, he had been readmitted to hospital after experiencing severe abdominal pain.

This time the consultant had found an abdominal lump the size of a grapefruit. Einstein underwent an operation known as a laparotomy to investigate the nature of the growth. It was found to be an aneurysm in the abdominal aorta. It was not removed but, because it was intact, it was decided that it should instead be regularly monitored and pain-relieving drugs were prescribed. Then, in 1950, doctors discovered that the aneurysm had grown. A short time later, Einstein put his name to his last will and testament.

A description of Einstein's will, written after his death for the use of the press, is still to be found among the tens of thousands of pieces of paper carefully filed in a vast collection of boxes in the basement of the library at the Princeton Institute of Advanced Study. It contains very few surprises.

By 1950, there were very few people in Einstein's life to whom he wished to leave his worldly possessions. Despite his enormous fame and almost universal admiration, he was a man who journeyed through life with very little baggage.

Einstein's first wife, Mileva, had died in 1948. She had remained in Switzerland living with their two sons, Eduard and Hans Albert. Eduard had been diagnosed as schizophrenic and had been looked after by Mileva until she had herself become too ill. At this point Eduard had been admitted to the Burghölzli psychiatric hospital in Zurich, where he died in 1965.

Hans Albert had left Switzerland before the outbreak of World War II and settled in America, becoming a professor of engineering at Berkeley, where he died in 1973. Einstein's sister Maja died in 1951. She had also settled in Princeton and, with Einstein's stepdaughter Margot and Helen Dukas, had made up the triumvirate of females who cared for the elderly scientist and kept him daily company in later years.

Einstein once said that he had been closer to Maja than to any other person in his life. The bonds forged in his very early childhood, before he had developed his intellectual image of the world, were perhaps the strongest. Maja was the sole survivor of Einstein's childhood and the last link with his European, familial roots. He must have deeply felt the loss of his only sister.

Einstein's will made provision for both sons, $15,000 going to Eduard and $10,000 to Hans Albert. However, the will had been changed after Maja's death and the majority of the estate was bequeathed to Margot and to his trusty secretary, Helen Dukas. The two of them also received a legacy of $20,000 each and they were made the sole trustees of all Einstein's literary property and rights until their deaths, when it was to be passed on to the Hebrew University in Jerusalem. Einstein's violin was to be left to his grandson Bernhard.

The last few years of Einstein's life were undoubtedly painful ones. He rarely complained and frequently insisted to doctors and nurses that he was 'feeling well', when in fact he probably was not. He remained in Mercer Street and was seen in the town of Princeton and at the institute less and less frequently. He continued to correspond with his colleagues around the globe and was still active with peace campaigns and his scientific work almost until the last day of his life. One of his final efforts involved a plan with Bertrand Russell to set up a conference of the world's leading scientists to discuss the implications of nuclear weapons. (This work continued after Einstein's death and grew into the influential series of meetings known as the Pugwash conferences.)

On 12 April 1955 Einstein collapsed at home. At the time Helen Dukas was on her own with Einstein and handled the crisis with typical calm and efficiency. The family doctor was immediately called and administered morphia injections. Hans Albert, notified that his father was critically ill, left California on the first flight to the east coast.

Einstein recovered a little over the following days. He was conscious and lucid, asking for paper and pens, writing letters, scribbling equations and talking animatedly with Margot and Helen Dukas. On Saturday morning, the 16th, Einstein took another turn for the worse and was in a great deal of pain. It was decided that he should be admitted to hospital.

According to his doctors, Einstein was very calm and clear-headed during his final days. He totally accepted that he was on the verge of death and it held no terrors for him. The aneurysm had ruptured and there was a leakage of blood from his aorta. Surgery was suggested but Einstein flatly refused to submit himself to this, preferring instead to let nature take its course.

Around 1.15 a.m. on 18 April the night nurse on duty was alerted to an irregularity in Einstein's breathing. With the help of another nurse, she raised the head of the bed to ease her patient's breathing, and heard him mumbling something. She leaned forwards to catch the words and realised that he was speaking in German, of which she was totally ignorant. Moments later Albert Einstein died.

Chapter Fourteen

Physics After Einstein

Einstein's major contributions to science ended, as we have seen, in 1926, with the advent of the new quantum theory that, ironically, he had 'godfathered'. In later years, he did still produce interesting scientific papers, especially (with hindsight) those on the gravitational lens effect and on cosmology. But he was never again to be involved with revolutionary developments that changed the face, not just of physics, but of the whole of science. So, when we talk of 'physics after Einstein', we do not just mean developments since his death in 1955, but developments since the establishment of the quantum theory, the last major scientific advance to which he contributed.

During the remainder of his life, Einstein's main scientific interest was in developing a unified field theory, an attempt to explain both gravity and electromagnetism in one mathematical package. He failed, and we do not intend to go into the details of the many failed attempts at producing a unified theory. Although it may seem at first sight that these were wasted years, scientifically speaking, from another perspective we can now see that Einstein was, in fact, a generation or two ahead of his time in this search for unification. For, throughout the 1980s and into the 1990s, the Holy Grail for which theoretical physicists have been searching is a theory of everything, or TOE, that will wrap up all of the known forces and fields of physics in one mathematical package.

The strong and the weak

There are two big differences between this modern search for unification and Einstein's earlier efforts. First, when Einstein set out on his quest in the early 1920s, the only two forces known to physics were gravity and electromagnetism, and it was natural to seek their unification. But we now know that there are two additional forces, which operate within the nucleus of an atom, and are known as the strong and the weak force. The strong force is what holds a nucleus together, in spite of the fact that it contains protons, each with positive charge, that ought to repel each other. The weak force is responsible for radioactive decay. Both these forces only have a short range, and cannot influence anything further away than the width of an atomic nucleus.

Gravity is the odd one out among these forces. Each of the other three can be described using quantum physics, with varying degrees of success. The quantum theory of electromagnetism, known as quantum electrodynamics, or QED, is the most successful scientific theory ever, in terms of the accuracy with which it describes the phenomena it is supposed to describe, such as the way electrons interact with one another. It is a relativistic quantum field theory, in the sense that it includes the requirements of the special theory of relativity and quantum physics; but it does not incorporate the general theory of relativity, which is the theory that describes gravity.

In the 1960s, physicists found how to include the weak force (or weak *field*) in the same package, and came up with an improved relativistic quantum field theory, called the electroweak interaction. It has proved harder to include the strong field in the package, but using the same basic structure as QED they have found a description of the strong field alone in a mathematical package known as quantum chromodynamics, or QCD. There are several different possible ways to combine electroweak theory with QCD, and the details of which approach is best are still being worked out, but such packages are, rather grandly, known as grand unified theories.

But although grand unified theories (or GUTs) include all three of the quantum fields that are known, they are not really that grand, because they do not include gravity. The ultimate aim – the TOE – is to combine the best grand unified theory with Einstein's general theory. That is, to combine the two great pillars of twentieth-century physics, gravity and quantum theory. The extreme difficulty of doing this results from two problems. First, gravity is a much weaker force than any of the other three, which makes it difficult to include it in the TOE on an equal footing.[1] Secondly, Einstein's general theory is what is called a 'classical' theory. It deals with smooth changes, not with quantum jumps. In that sense, it resembles the nineteenth-century version of electromagnetic theory, not the twentieth-century version, *quantum* electrodynamics. We may need a quantum theory of gravity first, before we can hope to combine it in one package with the three other quantum fields. And – as history has shown – the most fruitful approach is to combine the three quantum fields first, before trying to include gravity in the package.

The large and the small

There is another problem, related to the fact that gravity is such a weak force, compared with the other three. The effects of quantum gravity should become noticeable only under extreme conditions, where the gravitational field is very strong. 'Classical' theory, it turns out, should be an accurate description of gravity in the so-called 'weak field approximation' – which applies here and now. No experiment carried out on Earth can ever hope to measure directly reactions involving quantum aspects of gravity. Just about the only place where quantum gravity was important was in the big bang itself, the superdense fireball of energy in which the universe was born.

According to modern ideas, in the first split second of the birth of the universe, gravity was so strong (because all the matter in the universe was jammed together) that it was on an equal footing with the other three forces, and played a full part in the interactions between quanta. As the universe expanded, matter spread more thinly and gravity weakened, splitting off from the other fields to become a classical field. Later in the expansion, the other three fields each went their separate ways.

So the place to look for the influence of quantum gravity, and the place to 'test' the accuracy of a TOE, is in the big bang itself. In other words, use the theory to predict how the universe should have expanded outwards from the big bang, and compare those predictions with observations of the real universe.

That is why particle physicists (who deal with the very small) and cosmologists (who deal with the very large) are now learning about each other's ideas, and moving into each other's territory. Quantum physics and general relativity, both of them to a large extent Einstein's brainchildren, come together in the big bang.

A combination of the cosmological equations of the general theory and observations of the expanding universe has now established beyond reasonable doubt that the universe was indeed born in a superdense state, some 15 billion years ago, and has been expanding ever since. One of the key pieces of observational evidence is the so-called 'cosmic background radiation', a weak hiss of radio noise left over from the big bang itself and now filling all of space. The discovery in 1992 of minor variations in this background radiation (quickly dubbed 'ripples in time') that had been predicted by the theory served as the icing on the cake of big-bang cosmology, and persuaded remaining doubters that Einstein's equations really had been telling the truth back in 1917.

So the ultimate aim of modern physics, the search for a theory of everything, can be regarded as an attempt to unify the two great theories Einstein contributed so much to, the general theory and the quantum theory. And the test bed for those TOEs is the big bang, which was predicted by Einstein's equations, even though he did not, at first, believe what the equations were telling him. Physics after Einstein has, so far, by no means involved any revolutions overturning his ideas, but rather a continuing search to develop and unify the ideas of Einstein.

Before we tell you a little more about modern attempts at unification, though, there is one aspect of Einstein's own search which should be mentioned. His very first attempts at unifying gravity and electromagnetism, it turns out, have more in common with some aspects of current thinking than the later blind alleys up which he travelled.

Beyond the fifth dimension

One of the first attempts at constructing a unified theory, using a modification of the four-dimensional geometry of general relativity, was made by the mathematician Hermann Weyl, in 1918. The attempt was not, ultimately, to prove successful; but Weyl's approach caught the imagination of a young mathematician at the University of Königsberg (then in Germany, now Kaliningrad and part of Russia), called Theodor Kaluza. Early in 1919, Kaluza was struck by an inspired insight.[2] He realised that just as gravity could be explained in terms of the curvature of four-dimensional space-time, so electromagnetism could be explained in terms of the curvature of a *five*-dimensional 'space-time'. Writing the equations of the general theory of relativity in five dimensions gave the usual description of gravity, together with another set of field equations which are mathematically equivalent to Maxwell's equations of electromagnetism.

Kaluza immediately wrote to Einstein informing him of the breakthrough. In April 1919, Einstein replied: 'The idea of achieving [a unified theory] by means of a five-dimensional cylinder world never dawned on me . . . At first glance I like your idea enormously.'[3] But in spite of his initial enthusiasm, Einstein then proceeded to put up a stream of minor objections to Kaluza's presentation of the new idea, urging him repeatedly over the next few months to refine his arguments. It was only in 1921 that Einstein agreed to send the latest version of Kaluza's paper, with his own recommendation, to the Prussian Academy of Sciences. What Einstein recommended in 1921, the Prussian Academy was happy to publish, and Kaluza's idea duly appeared in print.[4]

Einstein's first published work on a unified field theory (a collaboration with Jakob Grommer) was written in 1922, and used the five-dimensional approach pioneered by Kaluza. He then turned to other matters (including other approaches to unified theory), before coming back to tackle the five-dimensional approach again in 1926 and 1927.

This time, he made a major step forward – although in fact someone else did it first. The obvious snag about Kaluza's original five-dimensional theory (apart from the rather important question of where the fifth dimension is, and what it represents) was that it did not take on board quantum theory – hardly surprisingly, since the complete version of quantum mechanics only emerged after 1926, seven years after Kaluza's flash of inspiration. Like the general theory itself, Kaluza's original five-dimensional idea was a classical theory. Einstein wrote two papers bringing together quantum theory and Kaluza's idea, and wrote to Hendrik Lorentz in February 1927: 'It appears that the union of gravitation and Maxwell's theory is achieved in a completely satisfactory way by the five-dimensional theory.'[5]

But there is a rather baffling aspect to this. As Abraham Pais points out in his scientific biography *Subtle is the Lord . . .*, before Einstein sent those two papers off for publication, he had seen a paper by the Swedish physicist Oskar Klein, which also incorporated the ideas of quantum physics into Kaluza's five-dimensional theory. (What Klein had done was to rewrite Schrödinger's equation with five variables instead of four, and show that the solution of this equation could be interpreted as describing particle-waves moving under the influence of both gravitational and electromagnetic fields.) Einstein had even written to his old collaborator Grommer, to point out the existence of Klein's paper. And at the request of his editor, Einstein added a reference to Klein's paper at the end of his own second paper on the same subject, with a note saying: 'Herr Mandel points out to me that the results communicated by me are not new. The entire content is found in the paper by O. Klein.' We echo Pais' comment: 'I fail to understand why he published his two notes in the first place.' Even assuming Einstein had carried out the work independently, by the time the papers were sent for publication he knew that he had been pre-empted by Klein.

The mystery will probably never be solved, but justice has been done historically, and the multidimensional approach to unification is known to this day as the 'Kaluza–Klein' theory. Einstein remained intermittently enthusiastic about the idea, publishing more papers on the subject in 1931–32 and between 1938 and 1941. But in between times he was equally (if not more) enthusiastic about various other attempts at unification, and by the end of his life the Kaluza–Klein theory seemed like just another dead end.

Curiously, the absence of the fifth dimension in the real world was *not* the reason why few people took the Kaluza–Klein approach seriously. There is a neat mathematical trick for getting rid of the fifth dimension. It is called 'compactification', and it works like this.

Think of a hosepipe. A hosepipe is a roughly cylindrical object, existing in three dimensions. It is made of a two-dimensional sheet of material, wrapped around in the third dimension to make a cylindrical pipe. But if you view a hosepipe from a long way away, it looks like a *one*-dimensional object, a curved line. It is as if two of the dimensions do not exist unless you look closely at the pipe. This is how mathematicians get rid of the fifth dimension. They say that space-time itself actually consists of little loops around the fifth dimension, so small that the compactified fifth dimension itself plays no part in everyday life. 'Points' in three-dimensional space are really little circles in four-dimensional space, according to Klein, with time as the remaining, fifth dimension. Since the circumference of one of these tiny 'hypercircles' is less than a billion-billionth the size of an atomic nucleus, it is no surprise that we cannot directly see this fifth dimension of space.

All that is simple enough, for the mathematicians. Where the Kaluza – Klein theory seemed to fall down, however, was with the discovery of the strong and weak nuclear forces. It seemed that physicists had a beautifully neat way to unify gravity with electromagnetism, but that the other forces could not be included in the package.

In the 1980s, however, the idea of the Kaluza – Klein approach to unification was revived. Some physicists realised that the strong and weak forces *could*, after all, be included in the package. But because these forces are more complicated than the simple inverse-square law forces of gravity and electromagnetism, adding in each of the 'new' forces to the picture requires not just one extra dimension of space, but several. In order to include all the observable features of all the four known forces of nature, you actually need a minimum of ten space dimensions, plus the usual one time dimension.

This makes the business of compactification more complex, and it is essential that all the extra dimensions of space are safely rolled up out of the way. The more dimensions you have, the more different ways there are to do the compactification, giving you different versions of a prospective TOE. But in one of the simplest examples, with a total of eleven dimensions, ordinary four-dimensional space-time is supplemented by seven dimensions of space rolled up into the seven-dimensional equivalent of a sphere.

Such ideas are still a long way from being developed into a fully worked-out theory of everything, and there is no definite proof that the Kaluza–Klein approach will indeed prove the right one in the end. But it is a salutary reminder, to anyone inclined to dismiss Einstein's efforts at unification as the rather cranky hobby of a once-great physicist who had long passed his prime, that one of the ideas that Einstein championed in the 1920s and 1930s is now, more than half a century later, proving one of the most fruitful areas of research for physicists seeking a theory of everything.

Stringing along

The central idea of this promising new approach to unification is that the conventional picture of particles such as electrons in terms of little points, with no extension in any direction, is replaced by a description in terms of objects which have extension in one dimension, like a line drawn on a piece of paper. But these 'lines' are very short indeed – it would take 100,000,000,000,000,000,000 of these 'strings' to stretch across the diameter of a proton.[6]

String theory was dreamed up in the early 1970s, as a description of a specific kind of interaction between particles, and at that time it was not thought to be a step on the road to unification. Mathematical physicists played with the equations of such tiny strings as much out of interest in the mathematics itself as through any hope of applying those equations to real problems in physics. But in the 1980s it turned out, to the surprise of many of the people involved in this work, that strings might have further-reaching implications than they had suspected.

The key discovery came when the mathematicians were attempting to describe the forces of physics in terms of little 'lines' and loops of string, vibrating and interacting with one another. On that picture, a piece of string with two ends can both rotate and vibrate, and properties such as electric charge can be thought of as tied to the end points of the strings. 'Particles' are described in terms of open strings, while the forces that operate between particles (such as electric forces) are carried by little loops of string. One particular kind of string loop, for example, represents the photon, the carrier of the electromagnetic force. The properties of 'particles' such as electrons are described, in such a theory, in terms of the vibrations of the little strings. And the interactions involving those particles, such as the way two electrons will be repelled from one another by the electric force, are explained in terms of the way lines and loops of string collide, merge briefly and break apart.

When the experts attempted to set up a quantum description of the forces of nature in terms of vibrating strings, they didn't bother trying to include gravity. By the 1980s, everybody knew that gravity was the most difficult force to quantise, and they expected that even if they were lucky enough to unify the other three forces within their description, it would be hard work squeezing gravity into the package. But to their surprise, it turns out that when string theory is set up to describe the three forces of nature that had already been given a quantum description, the equations automatically included closed loops of string, which emerged naturally from the calculations (rather like the way in which the speed of light emerges naturally from Maxwell's equations) and had not been 'put in' by the theorists. When they investigated the theoretical properties of those closed loops, by solving the appropriate equations, they discovered that they are gravitons, the carriers of the gravitational force. String theory seems to bring gravity into the unified package whether you ask for it or not.

This does not mean that Einstein's dream of unification has been fulfilled. There are still problems with string theory, and the biggest difficulty the physicists face is that they have no direct experimental evidence to go on. In the 1920s, when quantum theory was developed, experiments could be carried out on things like electrons to test the theory. But the smaller the particle you wish to probe in this way, the more energy you need in your particle accelerator; the experiments required to reveal the presence of strings directly would involve energies vastly greater than anything that can be achieved on Earth. Such energies existed only in the big bang in which the universe was born. So although string theory in general remains a promising line of attack on the problem of unification, exactly which version of string theory is best remains a mystery. It will be many years before we have anything like a genuine complete unified theory. (There is a bonus, though. If anyone ever solves that mystery, the chances are that they will be able to explain the big bang itself in the same package.)

But the reason why we have gone into so much detail in describing a theory which has not yet stood the test of time, and which seems so far removed from Einstein's own work, is this. String theory works only in ten dimensions (nine of space plus one of time). Like other attempts at unification, it requires just the kind of compactification of the extra dimensions invoked in Kaluza–Klein theories, and it can be regarded as a logical development from Einstein's own first attempts at unification, in the early 1920s. The path by which physicists have got from Kaluza's original idea to modern 'superstring' theory may have been circuitous, but Einstein himself would certainly have understood, and felt comfortable with, this modern version of unified field theory.

Even this, however, may seem somewhat esoteric to most people – a far cry from the explanation of why the sky is blue, how sugar dissolves in hot tea, how lasers and atom bombs work, the way sand is suspended in liquid cement, and the way electrons are knocked out of atoms by photons. Einstein's work has, as we have seen, had a direct and practical impact on our daily lives in the twentieth century. But perhaps the greatest mark of his scientific genius lies beyond the scope of our daily lives, beyond planet Earth, and beyond even the Solar System. Einstein's masterwork, the general theory of relativity, describes *the whole universe*, and explains the most energetic and intriguing phenomena found in the universe. A generation or more ahead of his time, he gave physicists the theoretical tools to describe the big bang, quasars, pulsars and black holes. We live in Einstein's universe, and the only way to end our account of his scientific achievements is by looking at his description of the universe in which we live.

Einstein's universe

The general theory of relativity remained essentially a plaything for mathematicians for almost half a century after Einstein presented his ideas to the Prussian Academy. Although the mathematicians were able to describe phenomena such as the bending of light in a gravitational lens, or the ultimate collapse of matter into a black hole, in which space-time is bent completely around on itself so that nothing can escape, hardly anyone thought that such extreme conditions could exist in the real universe. Even the theory of the big bang and the expanding universe, which now seems the greatest triumph of the general theory, was not really taken seriously until the middle of the 1960s.

Although astronomers had discovered the expansion of the universe at the end of the 1920s, and theorists had realised that this expansion of space itself was exactly predicted by the general theory, there was a sense of unreality about the notion that equations written down on a piece of paper here on Earth could really be describing events millions of light years away and billions of years ago.

In the 1940s and 1950s rival theories grew up to 'explain' the expansion of the universe. One school of thought held that Einstein's equations should be taken at face value, implying that long ago all the material of all the stars and galaxies in the universe was concentrated, along with all the shrunken space from between the stars and galaxies, in one point, from which it burst out in a big bang. The other school of thought held that the universe need not have started in such a dramatic fashion, even though it was expanding. In an infinitely large universe, they said, space could expand eternally, making the universe larger still, and new matter to make new stars and galaxies could appear out of nothing at all to fill in the gaps as old galaxies moved further apart. This 'continual creation' of matter was, they argued, no more implausible than the idea of *all* the matter being created at once in the big bang.

It was like a game of cosmic chess, in which theorists dreamed up new ideas, and observers found ways to test them, but nobody really believed, deep down inside, that any of the theoretical models necessarily contained *the* truth about the universe. Then, in the early 1960s, radio astronomers discovered the now famous cosmic background radiation, a hiss of noise at microwave frequencies that fills the entire universe. The impact was dramatic. This background radiation had actually been predicted by the big-bang theory, based on the simplest and most literal interpretation of Einstein's equations, back in the 1940s. It was interpreted in the 1960s as the echo of the big bang itself, and over the following thirty years ever more refined studies of the radiation confirmed in increasingly precise detail the accuracy of this description of the universe.[7]

Because radio waves travel at the speed of light, and these particular radio waves have been spreading out with the expansion of the universe, ever since the big bang itself, they can be interpreted not so much as messengers from a distant *place*, but from a distant *time* – a time just after the big bang, about 15 billion years ago.

Thanks to the explosion of interest in the big-bang theory caused by this discovery, in the 1970s and 1980s particle physicists realised that their theories, including ideas about unification, could be tested by using the big bang itself as their 'experiment', in the way that we have already mentioned. Gradually, this symbiosis between particle theorists and cosmologists led to an improved picture of the big bang, one which implies that in addition to all the bright stars and galaxies we can see in the universe there must be about a hundred times more 'dark matter', undetectable to our telescopes, but which played an important part in the formation of galaxies through its gravitational influence when the universe was young.

If this new cosmology was correct, the theorists calculated that there ought to be irregularities in the cosmic background radiation, which ought to look slightly brighter in some parts of the sky and slightly dimmer in others. This patchiness in the background radiation would correspond to the patchiness in the distribution of matter billions of years ago, when the radiation set out on its journey. These 'ripples' in the background radiation would mark the locations, long ago, of huge concentrations of dark matter as it began to pull together the material that would one day make up the galaxies. But if these ripples from the past – ripples in time – could not be detected, that would imply that, after all, the big-bang theory itself might be fatally flawed.

At the end of the 1980s, a satellite known as COBE (from Cosmic Background Explorer) was launched by NASA to study the background radiation with more precision than ever before. In 1992, the NASA team announced that they had discovered exactly the kind of ripples in time that the theory had predicted. It was headline news around the world – the combination of Einstein's general theory, the big-bang model, and the added ingredient of dark matter, had been vindicated. This was, and is, the most compelling evidence ever that the universe we live in is described by the equations of the general theory of relativity – that it is, indeed, Einstein's universe.

Appendix

My Credo

It is of course impossible to guage fully the significance of any one person's life against the total backdrop of human history, even if that person was Albert Einstein, for how should we judge a person?

When all is said and done, is a man or a woman the sum of their contributions to the world or is there more? Does it really – matter whether or not a person was good, sociable, honest, funny or handsome, or is it only what they achieved that matters?

Naturally, no-one can portray the total man, we can only collect glimpses into the subject's mind, piece together the evidence and sketch the outline of a life in an effort to represent the character and the essence of a person.

However, it is clear that on any level, Albert Einstein was a truly exceptional human being and rightly perceived as one of the greatest achievers, not just within the history of science but within the broader canvas of the human experience. He was both an intellectual giant and a man who possessed an instinctive morality which demanded that he try his utmost to make the world a better place. He was a man brimming over with paradoxes, a lover of humankind, but close to very few, a pacifist who advocated the creation of an atomic weapons programme and a man obsessed with a hatred of regimentation but beguiled with the strict beauty of mathematics.

To successive generations Albert Einstein has been the figurehead of twentieth-century science. He has become an icon of the advertising world and the mass media, a face instantly recognised around the globe by people of all ages and races. How surprising such a thing would have appeared to the young man who, at the turn of this century had only recently graduated from college and could not find a job.

Perhaps the closest we can come to an insight into the real personality of such a public figure as Albert Einstein is to read 'My Credo', a speech by Albert Einstein to the German League of Human Rights, Berlin, autumn 1932.[1]

Our situation on this earth seems strange. Every one of us appears here involuntarily and uninvited for a short stay, without knowing the whys and the wherefore. In our daily lives we only feel that man is here for the sake of others, for those whom we love and for many other beings whose fate is connected with our own.

I am often worried at the thought that my life is based to such a large extent on the work of my fellow human beings, and I am aware of my great indebtedness to them.

I do not believe in freedom, of the will. Schopenhauer's words: 'Man can do what he wants, but he cannot will what he wills' accompany me in all situations throughout my life and reconcile me with the actions of others, even if they are rather painful to me. This awareness of the lack of freedom of will preserves me from taking too seriously myself and my fellow men as acting and deciding individuals and from losing my temper.

I never coveted affluence and luxury and even despise them a good deal. My passion for social justice has often brought me into conflict with people, as did my aversion to any obligation and dependence I do not regard as absolutely necessary. I always have a high regard for the individual and have an insuperable distaste for violence and clubmanship. All these motives made me into a passionate pacifist and anti-militarist. I am against any nationalism, even in the guise of mere patriotism.

Privileges based on position and property have always seemed to me unjust and pernicious, as did any exaggerated personality cult. I am an adherent of the ideal of democracy, although I well know the weaknesses of the democratic form of government. Social equality and economic protection of the individual appeared to me always as the important communal aims of the state.

Although I am a typical loner in daily life, my consciousness of belonging to the invisible community of those who strive for truth, beauty and justice has preserved me from feeling isolated.

The most beautiful and deepest experience a man can have is the sense of the mysterious. It is the underlying principle of religion as well as all serious endeavour in art and science. He who never had this experience seems to me, if not dead, then at least blind. To sense that behind anything that can be experienced there is a something that our mind cannot grasp and whose beauty and sublimity reaches us only indirectly and as a feeble reflection, this is religiousness. In this sense I am religious. To me it suffices to wonder at these secrets and to attempt humbly to grasp with my mind a mere image of the lofty structure of all that there is.

Further Reading

Books marked with an asterisk include equations and are more technical (at least in parts) than the present book. The rest are accessible at about the same level as this book, or at a slightly higher technical level.

Mainly about Einstein

Ronald W. Clark, *Einstein: The life and times* (Hodder & Stoughton, London, 1973).
Helen Dukas and Banesh Hoffman (editors), *Albert Einstein: The human side* (Princeton University Press, New Jersey, 1979).
Kenji Sugimoto, *Albert Einstein: A photographic biography* (Schocken Books, New York, 1989).

Mainly about Einstein's work and/or its implications

Nigel Calder, *Einstein's Universe* (Viking, New York, 1979).
John Gribbin, *In Search of the Big Bang* (Bantam, New York, and Corgi, London, 1986).
John Gribbin, *In Search of the Edge of Time* (Bantam, London, and Harmony, New York, 1992).
John Gribbin, *In the Beginning* (Viking, London, and Little, Brown, New York, 1993).
Banesh Hoffmann, *Relativity and its Roots* (Scientific American/W. H. Freeman, New York, 1983).
*Julian Schwinger, *Einstein's Legacy* (Scientific American/W. H. Freeman, New York, 1986).
Clifford Will, *Was Einstein Right?* (Basic Books, New York, 1986).

Michael White and John Gribbin, *Stephen Hawking: A life in science* (Viking, London, and NAL, New York, 1992) deals with many of the implications of Einstein's general theory of relativity for the investigation of black holes and the big bang.

About Einstein's life and work

*Jeremy Bernstein, *Einstein* (Fontana, London, 1973).
The Born–Einstein Letters (Macmillan, London, 1971).
*A. P. French (editor), *Einstein: A centenary volume* (Harvard University Press, Cambridge, Massachusetts, 1979).
Banesh Hoffmann, *Albert Einstein: Creator and rebel* (Hart-Davis MacGibbon, London, 1972).
*Abraham Pais, *Subtle is the Lord* . . . (Oxford University Press, 1982).

Notes

Chapter One

1 Maja Winterler-Einstein, *Albert Einstein* (published in *The Collected Papers of Albert Einstein*, Vol. I, *The Early Years (1879–1902)*, Princeton University Press, New Jersey, 1987).
2 Ibid.
3 Anthony Storr, *The Dynamics of Creation* (Penguin, London, 1976), p. 86.
4 Albert Einstein (translated by Paul Arthur Schilpp), *Autobiographical Notes*. First appeared in Schilpp, *Albert Einstein – Philosopher-Scientist* (Evanston, Illinois, 1949).
5 The theorem devised about 500 BC which states that in a right-angled triangle, the square of the hypotenuse (the side opposite the right angle) is equal to the sum of the squares of the other two sides.
6 Antonia Vallentin, *Einstein: A Biography* (Weidenfeld and Nicolson, London, 1954).

Chapter Two

1 Galileo was born, incidentally, in the same year as William Shakespeare, 1564.
2 In case you are wondering, the first law of thermodynamics is even simpler, and states that work and heat are interchangeable, heat is a form of energy, and the total amount of energy in a closed system is always conserved.

Chapter Three

1 Ronald W. Clark, *Einstein: The life and times*, p. 49.
2 Albert Einstein, 'Mes Projets d'avenir' (original in the Staatsarchiv, Kanton Aargau).

3 Joseph Schwartz and Michael McGuinness, *Einstein for Beginners* (Writers and Readers Publishing, UK, 1979), p. 40.
4 Albert Einstein (translated by Paul Arthur Schilpp), *Autobiographical Notes*, p. 15.
5 Letter to H. Zangger, summer 1912.
6 *The Collected Papers of Albert Einstein*, Vol. I, *The Early Years (1879–1902)* (Princeton University Press, New Jersey, 1987), letter No. 58. Original in the Hebrew University, Jerusalem, Israel.
7 *Ibid*, letter No. 48. Original in the Hebrew University, Jerusalem, Israel.
8 *Ibid*, verse in album of Anna Schmid. Original in the Hebrew University, Jerusalem, Israel.
9 Paul Arthur Schilpp (editor), *Albert Einstein: Philosopher-scientist* (Tudor, New York, 1949), pp. 15–18.

Chapter Four

1 Pais, *Subtle is the Lord* . . ., p. 89.
2 Quoted by Born in P. A. Schilpp (editor), *Albert Einstein: Philosopher-scientist* (Tudor, New York, 1949), p. 46.
3 Thomas Young, *Miscellaneous Works* (reprinted by Johnson, New York, 1972), Vol. IX p. 461.
4 'Whiz' really is the operative word here. At 0 °C, the molecules in air are moving at several hundred metres per second, something that the pioneers of kinetic theory such as Kelvin fully appreciated, in the second half of the nineteenth century, from their studies of the way the pressure exerted by a gas changes when it is squeezed into a smaller volume.

Chapter Five

1 Letter to Wisseler, 24 August 1948, now in the National Library, Bern.
2 *The Collected Papers of Albert Einstein*, Vol. I, *The Early Years (1879–1902)* (Princeton University Press, New Jersey), document No. 114.
3 Ibid., document No. 128, 17 December 1901.
4 Ibid., document No. 134, 4 February 1902.
5 *Einstein, Lettres à Maurice Solovine* (Gauthier-Villars, Paris, 1956), p. vi.

6 Philipp Frank, *Einstein: His Life and Times* (Da Capo series in science edition, 1989), p. 75.

Chapter Six

1 Quoted, for example, by Pais, p. 88.
2 Einstein actually wrote *six* scientific papers in 1905, all of which were published.
3 *Annalen der Physik*, Vol. XVII, 1905, p. 549.
4 *Ibid*., p. 132.
5 If you want a more detailed account of the birth of quantum theory, see *In Search of Schrödinger's Cat*, by John Gribbin (Bantam, London, 1984).
6 It appeared, of course, in the *Annalen der Physik* (Vol. XIV, 1904 p. 354).
7 *Reviews of Modern Physics*, Vol. XXI, 1949, p. 343.
8 *Annalen der Physik*, Vol. XVII, 1905, p. 891.
9 Quoted, for example, by Pais, p. 152.
10 This was the subject of yet another paper, a very short one, published by Einstein in the *Annalen* soon after the June 1905 paper announcing the special theory.
11 It was also, as we shall see in Chapter Eight, to provide the key insight in the development of Einstein's masterwork, the general theory of relativity.

Chapter Seven

1 Letter to M. Besso, 17 November 1909, in P. Speziali (editor), *Albert Einstein–Michelangelo Besso Correspondence, 1903–1955* (Hermann, Paris, 1972).
2 Philipp Frank, *Einstein: His Life and Times* (Da Capo, 1989), p. 101.
3 Ibid., p. 78.
4 Letter to Besso, 13 May 1911, in *Albert Einstein–Michelangelo Besso Correspondence*, p. 19.
5 Madame Curie to Professor Weiss, 17 November 1911 (ETH).
6 Poincaré to ETH, undated (ETH).
7 Original in the Hebrew University, Jerusalem, Israel.
8 Letter to Heinrich Zangger, spring 1915.
9 John Plesch, *Janos: The Story of a Doctor* (London Books, 1947), p. 206.
10 Letter to Carl Seelig, 5 May 1952.

Chapter Eight

1 Quoted by Pais, p. 178; the italics are Einstein's.
2 The talk is known as the 'Kyoto address'; it is reported by J. Ishiwara in *Einstein Koen-Roku* (Tokyo-Tosho, Tokyo, 1977).
3 See John Gribbin, *In Search of the Edge of Time*.
4 Preussische Akademie der Wissenschaften (Proceedings of the Prussian Academy of Sciences), Part I, 1917, p. 142.
5 See Clifford Will, *Was Einstein Right*? p. 162.
6 The irony is that the *special* theory of relativity really does overturn Newtonian ideas about mechanics, changing such fundamental ideas as our understanding of mass and how to add up velocities. But in 1905, this made scarcely a ripple in the world outside the scientific community.

Chapter Nine

1 A. N. Whitehead, *Science and the Modern World* (London Books) p. 13.
2 The *Times*, 8 November 1919.
3 *Ibid.*
4 A. Moszkowski, *Einstein the Searcher* (London Books) p. 13–14.
5 The *Times*, 28 November 1919.
6 The *Nation*, 27 December 1919, p. 819.
7 Virginia Woolf, *Roger Fry* (Harcourt, Brace and Co., New York, 1940), p. 153.
8 Letter to H. Zangger, undated, March 1920.
9 Philipp Frank, *Einstein: His life and times* (Da Capo, 1989), p. 173.
10 From the Haldane correspondence in the National Library of Scotland, p. 1052.
11 *Ibid.*, p. 189.

Chapter Ten

1 The particles used were alpha particles, which carry two units of positive charge and are produced naturally by some radioactive atoms. We now know that an alpha particle is the same as the nucleus of a helium atom, and is composed of two protons and two neutrons bound together by the so-called 'strong nuclear force'.
2 The resulting paper was published the following year, under

the title 'On the quantum theory of radiation' (*Physikalische Zeitschrift*, Vol. xviii, 1917, p. 121.

3 The citations are discussed by Tony Cawkell and Eugene Garfield in their contribution to the book *Einstein: The first hundred years*, edited by M. Goldsmith, A. Mackay and J. Woodhuysen (Pergamon Press, Oxford, 1980).

4 Another way of thinking about this is to say that there is no distinction between a photon and an antiphoton.

5 It would be an exaggeration to say that he did this by trial and error, but it certainly helped that he knew the answer – Planck's equation – that he was looking for.

6 Preussische Akademic der Wissenschaften (Proceedings of the Prussian Academy of Sciences) Part I, 1925, p. 3.

7 Pais, *Subtle is the Lord* . . ., p. 438.

8 For the details, see John Gribbin, *In Search of Schrödinger's Cat*.

9 When Einstein, who was an atheist, used the expression 'God' or 'the Old One', he meant what most people would mean by the expression 'Mother Nature'.

Chapter Eleven

1 Curie–Einstein letters, July (date unclear) 1922 (Institut de Physique Nucleaire, Paris).

2 Banesh Hoffmann, *Albert Einstein: Creator and rebel*, p. 155.

3 *Berliner Lokal-Anzeiger*, March 1933.

4 Statement to Associated Press, 3 December 1932.

5 Letter to Alfred Nahon, 20 July 1933 (*La Patrie humaine*, 18 August 1933).

6 Hoffmann, p. 157.

Chapter Twelve

1 Preussische Alcademie der Wissenschaften (Proceedings of the Prussian Academy of Science), Part I, 1925, p. 3.

2 Details of all this are discussed in *In Search of Schrödinger's Cat*, by John Gribbin.

3 The key feature of quantum jumping that so worried Schrödinger, Einstein and others is that it is *discontinuous*. An electron, for example, 'jumps' from point A to point B instantly, without occupying any of the space in between. It leaves one energy level and reappears in another without any time interval in

which it can make the transition. Popular usage sometimes employs 'quantum leap' for a *large* step, but this misses the point. A quantum leap is much more likely to be a very small change, like the jump of an electron from one energy level to the next within an atom. Indeed, the *smallest possible* change is always a quantum leap, and, like absorption and emission of photons, obeys the rules of probability and chance. So when advertisers tell you that this year's model represents 'a quantum leap' over last year's, they are telling you (possibly truthfully) that they have made the smallest possible change, and that it has been made at random.

4 Cat lovers may be reassured to know that nobody has ever carried out this experiment; it is purely a hypothetical example.

5 The letter was dated 4 December 1926; see *The Born–Einstein Letters*, p. 90. This is the source of the famous Einstein 'quote' (actually a paraphrase) 'God does not play dice'. In *Subtle is the Lord . . .* (p. 443), Pais quotes from the same letter, with a slightly different translation of the original German.

6 *Physical Review*, Vol. 47, 1935, p. 777.

7 P. C. W. Davies and J. R. Brown (editors), *The Ghost in the Atom* (Cambridge University Press, 1986).

8 Heinz Pagels, *The Cosmic Code* (Simon & Schuster, New York, 1982).

Chapter Thirteen

1 Postal telegram, 28 September 1933 (Princeton archive).

2 Undated, *Born–Einstein Letters*, p. 128.

3 Albert Einstein, 'Atomic War or Peace', *Atlantic Monthly*, November 1945.

4 Original in the Hebrew University, Jerusalem, Israel.

5 Quoted in Otto Nathan and Heinz Norden (editors), *Einstein on Peace* (Simon & Schuster, New York, 1963), p. 572.

Chapter Fourteen

1 If you find it hard to think of gravity as a *weak* force, remember that it takes the gravity of the entire planet Earth to pull an apple to the ground. Yet when the same apple is hanging from the tree by its stalk, the force that holds it up, overcoming gravity, is the electromagnetic interaction of a few atoms in the stalk. Gravity only seems like a strong force to us because it is a *long-range*

force, so that we feel the gravitational effect of all the atoms in the Earth put together. But a child of two can overcome the Earth's gravity and stand upright. The child is, literally, stronger than the entire Earth's gravitational pull.

2 The story is told in more detail in *In Search of the Big Bang*, by John Gribbin.

3 Letter dated 21 April 1919, quoted by Pais, p. 330.

4 Preussische Alcademie der Wissenschaften (Proceedings of the Prussian Academy of Science), 1921, p. 966.

5 Quoted by Pais, p. 333.

6 Protons are not thought to be fundamental 'particles', but are made up of smaller entities, known as quarks.

7 Over those same decades, astronomers discovered many violent and energetic objects, such as quasars, in the universe; with their increasing confidence in Einstein's equations as a literal description of what is going on 'out there', these have been satisfactorily explained in terms of black holes, within the framework of the general theory.

Appendix

1 Original in the Hebrew University, Jerusalem, Israel.

Index

Albert Einstein is referred to in this index as A.E.